OXFORD REVISION GUIDES

A Level

Advanced

GEOGRAPHY

through diagrams

Garrett Nagle

Kris Spencer

OXFORD
UNIVERSITY PRESS

OXFORD

UNIVERSITY PRESS

Great Clarendon Street, Oxford OX2 6DP

Oxford University Press is a department of the University of Oxford.
It furthers the University's objective of excellence in research, scholarship,
and education by publishing worldwide in

Oxford New York

Athens Auckland Bangkok Bogotá Buenos Aires Calcutta
Cape Town Chennai Dar es Salaam Delhi Florence Hong Kong Istanbul
Karachi Kuala Lumpur Madrid Melbourne Mexico City Mumbai
Nairobi Paris São Paulo Singapore Taipei Tokyo Toronto Warsaw

with associated companies in Berlin Ibadan

Oxford is a registered trade mark of Oxford University Press
in the UK and in certain other countries

To Angela

ISBN 0 19 913404 9 Student's Edition
First published 1997
New Edition 1998
Reprinted in 1999 (twice)

ISBN 0 19 913405 7 Bookshop Edition
First published 1997
New Edition 1998
Reprinted in 1999 (twice)

Every effort has been made to trace and contact copyright holders of
material reproduced in this book. Any omissions will be rectified in
subsequent printings if notice is given to the publisher.

Typesetting, artwork, and design by Hardlines, Charlbury, Oxford

Printed in Great Britain

CONTENTS

The structure of the Earth

WORKING FROM THE CENTRE OUTWARDS

1 **Core:** Solid, consists of iron and nickel. Density about 13.6 g cm^{-3}, approximately five times more dense than surface rocks.

2 **Outer core:** Liquid, consists largely of iron. Density about 10–12 g cm^{-3}. It is believed that the earth's magnetic field is generated by movements in the liquid outer core.

3 **Mantle:** Solid, consists of lower density material (4–5 g cm^{-3}) known as peridotite, a material composed of silicate minerals. Approximately 2900 km thick, may be divided into two subdivisions, the upper and lower mantle.

4 **Crust:** Solid, divided into two different types, continental and oceanic crust. Depth varies from 10 to 35 km, density about 3 g cm^{-3}.

Continental crust is largely composed of granite and is sometimes referred to as *sial* due to the volume of *si*lica and *al*uminium in its make up. Continental crust is less dense than the basaltic oceanic crust (also known as *sima*, because of the *si*lica and *ma*gnesium in its make up) and also considerably thicker. It appears that the oceanic crust plunges down beneath the continental crust; the division between the two layers is known as the Conrad Discontinuity.

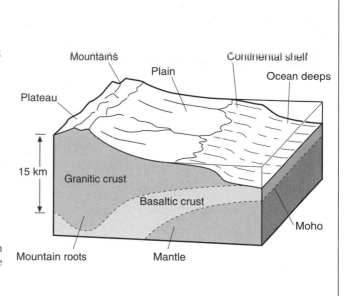

EVIDENCE RELATING TO THE INTERNAL STRUCTURE OF THE EARTH

Two types of waves produced:

- surface waves from epicentre

- body waves from focus - body waves can be divided into primary (P) waves and secondary (S) waves

P waves travel in the Earth's crust at an average speed of 6.1 km/s. S waves travel at 4.1 km/s. S waves cannot travel through liquids.

P waves take 14 minutes to reach here, S waves take 25 minutes. At D there are no S waves, and the P waves arrive much more slowly than would be suggested by their normal speed through the mantle. The lack of S waves and the slowness of P waves suggests that the outer core is liquid.

The **lithosphere** is a relatively inflexible and bouyant layer. It is this layer which floats on the material underneath and as it moves carries the continents - the tectonic plates - that produce *continental drift*.

The **asthenosphere** is the layer below the lithosphere. Seismic waves decrease with distance through this region. This is possibly due to a state of flux caused by the high temperature.

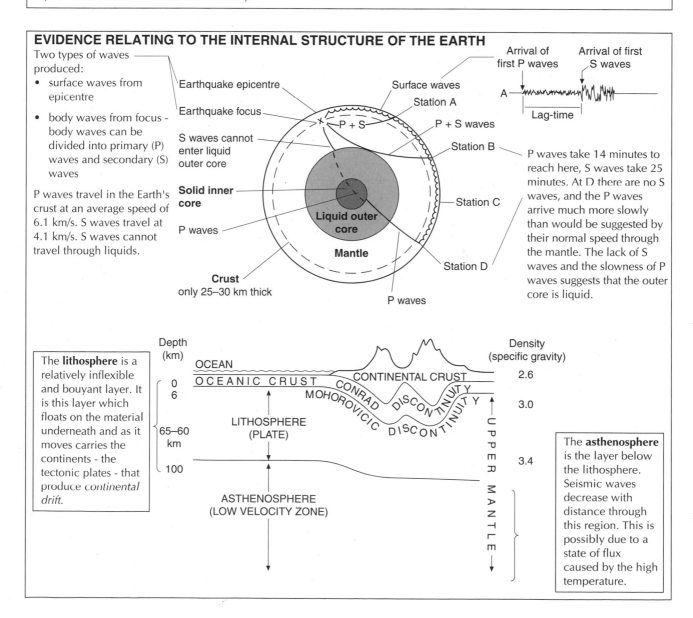

Processes at plate margins

LANDFORMS AND PLATE TECTONICS

The lithosphere is divided into a number of large and small rigid plates. There are three types of boundary:

(i) divergent - where plates are moving apart at ocean ridges or continental rifts

(ii) convergent - where plates are moving together and one plate is forced beneath another forming ocean trenches

(iii) transform or transcurrent - where plates are moving past each other and are neither constructive nor destructive

Diverging plates spread apart, splitting the crust. This is followed by the formation of new crust. They are therefore CONSTRUCTIVE. Converging plates involve major mountain building and subduction of the crust. They are known as DESTRUCTIVE.

WHAT HAPPENS AT CONSTRUCTIVE PLATE MARGINS?

Theories of plate motions and ocean ridges

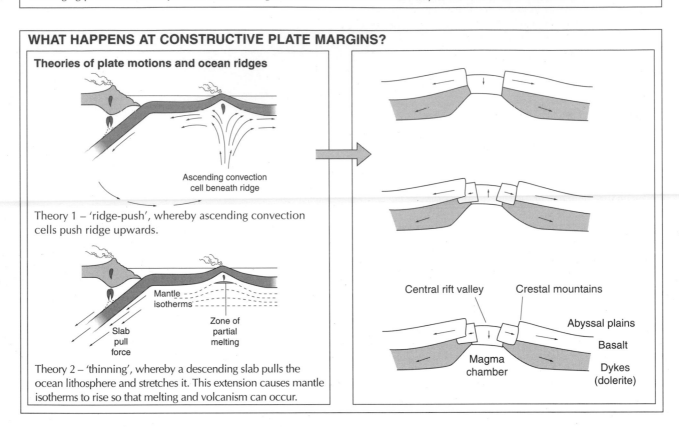

Ascending convection cell beneath ridge

Theory 1 – 'ridge-push', whereby ascending convection cells push ridge upwards.

Mantle isotherms

Slab pull force

Zone of partial melting

Theory 2 – 'thinning', whereby a descending slab pulls the ocean lithosphere and stretches it. This extension causes mantle isotherms to rise so that melting and volcanism can occur.

Central rift valley — Crestal mountains

Abyssal plains

Basalt

Magma chamber

Dykes (dolerite)

WHAT HAPPENS AT DESTRUCTIVE PLATE MARGINS?

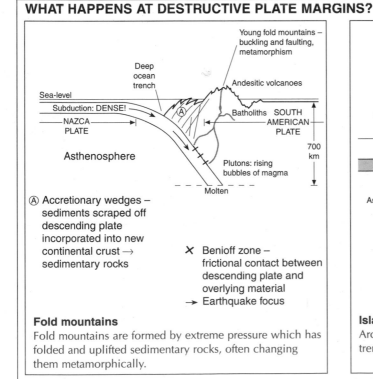

Deep ocean trench

Young fold mountains – buckling and faulting, metamorphism

Sea-level

Subduction: DENSE!

Andesitic volcanoes

NAZCA PLATE

Batholiths   SOUTH AMERICAN PLATE

Asthenosphere

700 km

Plutons: rising bubbles of magma

Molten

(A) Accretionary wedges – sediments scraped off descending plate incorporated into new continental crust → sedimentary rocks

✕ Benioff zone – frictional contact between descending plate and overlying material
→ Earthquake focus

Fold mountains

Fold mountains are formed by extreme pressure which has folded and uplifted sedimentary rocks, often changing them metamorphically.

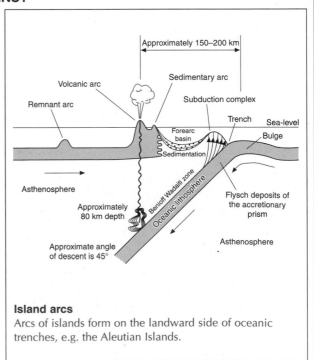

Approximately 150–200 km

Volcanic arc

Sedimentary arc

Remnant arc

Subduction complex

Trench   Sea-level

Forearc basin

Bulge

Sedimentation

Benioff Wadaiti zone

Oceanic lithosphere

Asthenosphere

Approximately 80 km depth

Flysch deposits of the accretionary prism

Approximate angle of descent is 45°

Asthenosphere

Island arcs

Arcs of islands form on the landward side of oceanic trenches, e.g. the Aleutian Islands.

Earthquakes

CAUSES

Earthquakes occur when normal movements of the crust are concentrated into a single shock or a series of sudden shocks. Aftershocks occur later as stresses are redistributed. The sequence is as follows:

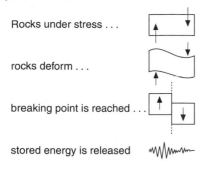

Rocks under stress . . .

rocks deform . . .

breaking point is reached . . .

stored energy is released

EARTHQUAKE DAMAGE

Tsunamis and mudslides	Type of building	Magnitude
Population density	**Factors affecting damage**	Season
Time of day	Distance from epicentre and depth of focus	Type of ground sediment

SHOCK WAVES

Waves associated with the focus

P waves: fast/compression
S waves: slower/distortion } travel through the interior

Waves associated with the epicentre

Love and Rayleigh waves which travel on the surface and cause the damage.

PREDICTION

- Crustal movement.
- Historical evidence.
- Seismic activity.
- Minor quakes before 'The Big One'.
- Change in properties of ocean crust.
- Gas omissions from ground.
- Changes in electrical conductivity.

CASE STUDY: THE KOBE EARTHQUAKE

Details: 17th January 1995, killed over 5000 people, injured over 30,000, and made almost 750,000 homeless.

Causes: Philippine plate is being subducted beneath the Eurasian plate. Kobe is situated in a geographically complex area near the northern tip of the Philippine plate.

Secondary factors: Rain and strong winds increased landslide risk; damp, unhygienic conditions encouraged disease; fires, broken glass, broken water pipes, and lack of insurance meant that many lost their livelihood.

HUMAN IMPACT

- Mining - gold-mining in the Witwatersand area of South Africa has been blamed for frequent seismic activity because of changed rock stress.
- Reservoirs - previously an area free from tectonic tremors, the states of Nevada and Arizona in the USA experienced over 100 tremors in 1937 following the construction of the Hoover Dam and the creation of Lake Mead due to seepage.

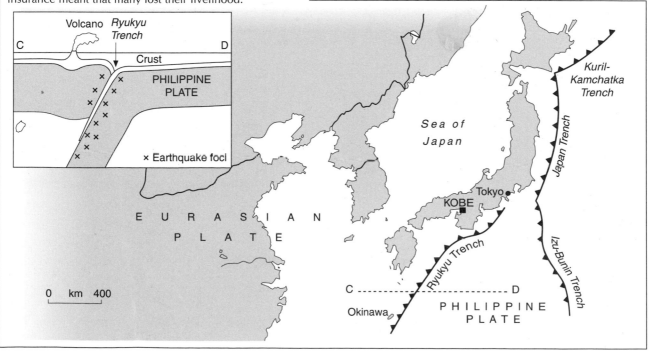

Volcanoes

FACTORS AFFECTING VOLCANIC LANDSCAPES

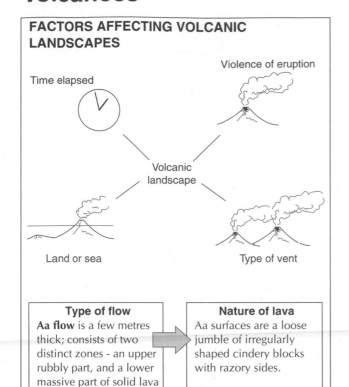

Time elapsed

Violence of eruption

Volcanic landscape

Land or sea

Type of vent

Type of flow	Nature of lava
Aa flow is a few metres thick; consists of two distinct zones - an upper rubbly part, and a lower massive part of solid lava which cools slowly.	Aa surfaces are a loose jumble of irregularly shaped cindery blocks with razory sides.
Pahoehoe flow is the least viscous of all lavas; rates of advance can be slow; cool surface, flow occurs underneath.	Pahoehoe surfaces can be smooth and glossy but may also have cavities; surface may also be crumpled with channels.

CASE STUDY: MOUNT PINATUBO

Details: 9th June 1991; eruption after 600 years; between 12th and 15th June ash and rock was scattered over a radius of 100 km; killed 350 people and made 200,000 people homeless, largely due to mudslides.

Effects: Mudslides covered 50,000 ha of cropland and destroyed 200,000 homes; 600,000 people lost their jobs.

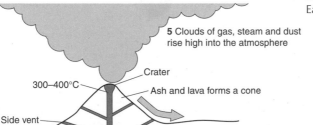

5 Clouds of gas, steam and dust rise high into the atmosphere

Crater

300–400°C

Ash and lava forms a cone

Side vent

3 Build-up of pressure over hundreds of years finally released in volcanic eruption

4 Volcanic ash, a mixture of gas, ash and molten rock, flows down mountain at 100 km per hour or more

Magma chamber 650–1200°C

Eurasian plate

2 Rocks of Philippine plate melt in high temperatures below Earth's crust, creating liquid magma which is forced up through cracks

Philippine plate

1 Thin Philippine plate driven beneath thicker Eurasian plate by continental drift

TYPES OF VOLCANIC CONES

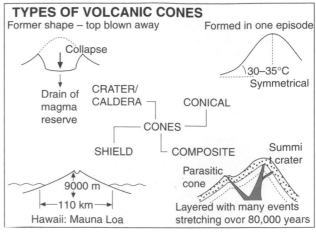

Former shape – top blown away

Collapse

Drain of magma reserve

CRATER/ CALDERA

Formed in one episode

30–35°C Symmetrical

CONICAL

CONES

SHIELD

COMPOSITE

9000 m

110 km

Hawaii: Mauna Loa

Summit crater

Parasitic cone

Layered with many events stretching over 80,000 years

TYPES OF VOLCANIC ERUPTIONS

HAWAIIAN TYPE Runny basaltic lava which travels down sides in lava flows. Gases escape easily.

STROMBOLIAN TYPE Frequent gas explosions which blast fragments of fairly runny lava into air, to form cone.

VULCANIAN TYPE Violent gas explosions blast out plugs of sticky or cooled lava. Fragments build up cone of ash and pumice.

VESUVIAN (SUB-PLINIAN) TYPE Very powerful blasts of gas push ash clouds high into sky. Lava flows also occur. Ash falls to cover surrounding area.

PLINIAN TYPE Gas rushes up through sticky lava and blasts ash and fragments into sky in huge explosion. Gas clouds and lava can also rush down slopes. Part of volcano may be blasted away during eruption.

Causes:

Earthquake 16th July 1990 (7.7 on Richter Scale; 1600 dead)

Basalt from the upper mantle squeezed into the magma chamber of the dormant volcano

Basalt reactivated viscous lava and created gas-charged magma (andesite)

This rose towards the surface causing volcano to bulge

Pressure blasted away the dome spewing 20 million tonnes of material into the atmosphere

Weathering

Weathering is the decomposition and disintegration of rocks in situ. Decomposition refers to the chemical process and creates altered rock substances whereas disintegration or mechanical weathering produces smaller, angular fragments of the same rock. Weathering is important for landscape evolution as it breaks down rock and facilitates erosion and transport.

MECHANICAL (PHYSICAL) WEATHERING

There are four main types of mechanical weathering: freeze-thaw (ice crystal growth), salt crystal growth, disintegration, and pressure release.

Freeze-thaw occurs when water in joints and cracks freezes at $0°C$ and expands by 10% and exerts pressure up to $2100 kg/cm^2$. Rocks can only withstand a maximum pressure of about $500 kg/cm^2$. It is most effective in environments where moisture is plentiful and there are frequent fluctuations above and below freezing point, e.g. periglacial and alpine regions.

Salt crystal growth occurs in two main ways: first, in areas where temperatures fluctuate around $26-28°C$, sodium sulphate (Na_2SO_4) and sodium carbonate (Na_2CO_3) expand by 300%; second, when water evaporates, salt crystals may be left behind to attack the structure. Both mechanisms are frequent in hot desert regions.

Disintegration is found in hot desert areas where there is a large diurnal temperature range. Rocks heat up and expand by day and cool and contract by night. As rock is a poor conductor of heat, stresses occur only in the outer layers and cause peeling or *exfoliation* to occur. Griggs (1936) showed that moisture is essential for this to happen.

Pressure release is the process whereby overlying rocks are removed by erosion thereby causing underlying ones to expand and fracture parallel to the surface. The removal of a great weight, such as a glacier, has the same effect.

CHEMICAL WEATHERING

There are four main types of chemical weathering: carbonation-solution, hydrolysis, hydration, and oxidation.

Carbonation-solution occurs on rocks containing calcium carbonate, e.g. chalk and limestone. Rainfall and dissolved carbon dioxide forms a weak carbonic acid. (Organic acids also acidify water.) Calcium carbonate reacts with an acid water and forms calcium bicarbonate, or calcium hydrogen carbonate, which is soluble and removed by percolating water.

Hydrolysis occurs on rocks containing orthoclase feldspar, e.g. granite. Orthoclase reacts with acid water and forms kaolinite (or kaolin or china clay), silicic acid, and potassium hydroxyl. The acid and hydroxyl are removed in the solution leaving china clay behind as the end product. Other minerals in the granite, such as quartz and mica, remain in the kaolin.

Hydration is the process whereby certain minerals absorb water, expand, and change, e.g. anhydrate becomes gypsum.

Oxidation occurs when iron compounds react with oxygen to produce a reddish brown coating.

LIMESTONE WEATHERING

Factors controlling the amount and rate of limestone solution

1 The amount of carbon dioxide (CO_2), which is controlled by:
   - the amount of atmospheric CO_2
   - the amount of CO_2 in the soil and groundwater
   - the amount of CO_2 in caves and caverns
   - temperature - CO_2 is more soluble at low temperatures
2 The amount of water in contact with the limestone.
3 Water temperature.
4 The turbulence of the water.
5 The presence of organic acids.
6 The presence of lead, iron sulphides, sodium, or potassium in the water.

Rates of limestone solution

Rates of solution vary. In the Burren, western Ireland, the average depth of solution is 8 cm on bare ground, 10 cm under soil, and 22 cm underground. This has taken place over the last 10,000 years, representing an average of 15 cm in 10,000 years or 0.0152 mm per year.

Accelerated solution

Accelerated solution occurs under certain conditions:

- Impermeable rocks adjoin limestone - waters from non-karstic areas have aggressive waters and will cause above average rates of solution.
- Alluvial corrosion - intense solution takes place by water which passes through alluvium and morainic sands and gravels.
- Corrosion by mixture - this occurs when waters of different hardness mix.
- At the margins of snow and ice fields - snow meltwater is able to dissolve more limestone than rain-water.
- Limestone denudation increases as annual rainfall and runoff increase.
- Limestone weathers more quickly under soil cover than on bare surfaces.

Human activity has more impacts on the nature and rate of limestone denudation:
- the burning of fossil fuels and deforestation has increased atmospheric levels of carbon dioxide and thus the weathering of limestone might accelerate
- acid rain is increasing levels of acidity (sulphur dioxide and nitrogen oxides) in rain-water
- agriculture and forestry are affecting soil acidity and carbon dioxide levels

Controls on weathering

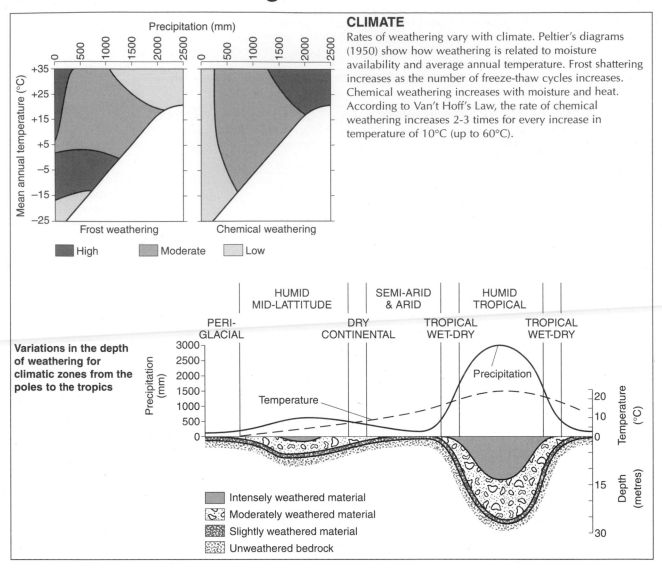

CLIMATE

Rates of weathering vary with climate. Peltier's diagrams (1950) show how weathering is related to moisture availability and average annual temperature. Frost shattering increases as the number of freeze-thaw cycles increases. Chemical weathering increases with moisture and heat. According to Van't Hoff's Law, the rate of chemical weathering increases 2-3 times for every increase in temperature of 10°C (up to 60°C).

Variations in the depth of weathering for climatic zones from the poles to the tropics

GEOLOGY

Rock type influences the rate and type of weathering in many ways due to:

- chemical composition

- the nature of cements in sedimentary rock

- joints and bedding planes

For example, limestone consists of calcium carbonate and is therefore susceptible to carbonation-solution, whereas granite with orthoclase feldspar is prone to hydrolysis. In sedimentary rocks, the nature of the cement is crucial: iron-oxide based cements are prone to oxidation whereas quartz cements are very resistant.

The importance of joints and bedding planes: the formation of tors

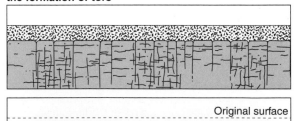

Breakdown of rock along joints and bedding planes

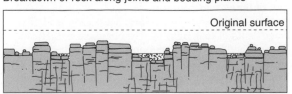

Removal of weathered material to expose tors, e.g. Hay Tor, Yes Tor on Dartmoor

Slopes

Slopes can be defined as any part of the solid land surface. They can be **sub-aerial** (exposed) or **sub-marine** (underwater), **aggradational** (depositional), **degrational** (eroded), or **transportational**, or any mixture of these. Geographers study the **hillslope**, which is the area between the **watershed** (or drainage basin divide) and the **base**, that may or may not contain a river or stream. Slope **form** refers to the shape of the slope in cross-section; slope **processes** the activities acting on the slopes; and slope **evolution** the development of slopes over time.

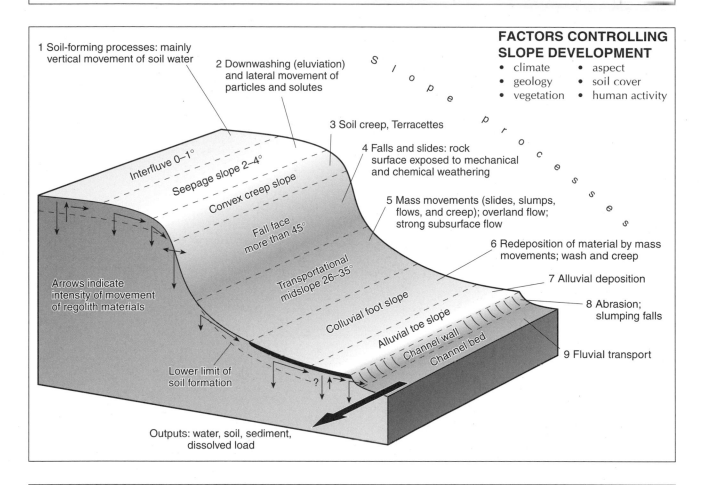

FACTORS CONTROLLING SLOPE DEVELOPMENT
- climate
- geology
- vegetation
- aspect
- soil cover
- human activity

1 Soil-forming processes: mainly vertical movement of soil water

2 Downwashing (eluviation) and lateral movement of particles and solutes

3 Soil creep, Terracettes

4 Falls and slides: rock surface exposed to mechanical and chemical weathering

5 Mass movements (slides, slumps, flows, and creep); overland flow; strong subsurface flow

6 Redeposition of material by mass movements; wash and creep

7 Alluvial deposition

8 Abrasion; slumping falls

9 Fluvial transport

Interfluve 0–1°

Seepage slope 2–4°

Convex creep slope

Fall face more than 45°

Transportational midslope 26–35°

Colluvial foot slope

Alluvial toe slope

Channel wall

Channel bed

Arrows indicate intensity of movement of regolith materials

Lower limit of soil formation

Outputs: water, soil, sediment, dissolved load

Slope processes

SLOPE PROCESSES: SOIL CREEP

Individual soil particles are pushed or heaved to the surface by wetting, heating, or freezing of water. They move at right angles to the surface (2) as it is the zone of least resistance. They fall (5) under the influence of gravity once the particles have dried, cooled, or the water has thawed. Net movement is downslope (6). Rates are slow - 1 mm/year in the UK and up to 5 mm/year in the tropical rainforest. They form **terracettes**, e.g. those at Maiden Castle, Dorset, and The Manger, Vale of the White Horse.

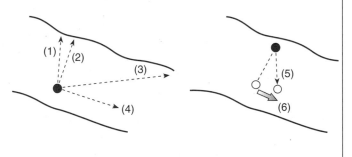

SLOPE PROCESSES: RAIN-SPLASH EROSION

On flat surfaces (A) raindrops compact the soil and dislodge particles equally in all directions. On steep slopes (B) the downward component (b) is more effective than the upward motion (a) due to gravity. Hence erosion downslope increases with slope angle.

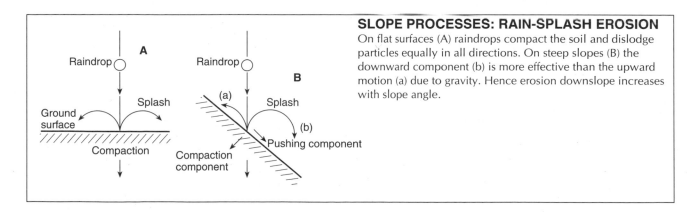

A
Raindrop
Splash
Ground surface
Compaction

B
Raindrop
(a)
Splash
(b)
Pushing component
Compaction component

Slope controls

GEOLOGY

Rock type affects slopes through its strength, dip, and orientation of joints and bedding planes.

Dip of rock

(i) Steeply dipping, e.g. Hogsback near Guildford

(ii) Gently dipping, e.g. North Downs

(iii) Horizontal strata, e.g. Salisbury Plain

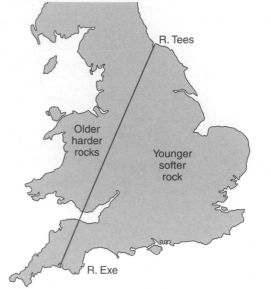

The **Tees-Exe line** is an imaginary line running from the River Tees to the River Exe. It divides Britain into hard and soft rock. To the north and west are old, hard, resistant rocks, e.g. granite, basalt, and carboniferous limestone, forming upland rugged areas. To the south and east are younger softer rocks, such as chalk and clay, forming more subdued low-lying landscapes.

CLIMATE

L.C. Peltier's classification of morphogenetic regions (1950)			
	Annual temp. (°C)	Annual rainfall(mm)	Processes
Glacial	−20 to −5	0 - 1100	Ice erosion Nivation Wind action
Periglacial	−15 to 0	125 - 1300	Mass movement Wind action Weak water action
Boreal	−10 to 5	250 - 1500	Moderate frost action Slight wind action Moderate water action
Maritime	5 to 20	1250 - 1500	Mass movement Running water
Selva (rainforest)	15 to 30	1400 - 2250	Mass movement Slight slope wash
Moderate (temperate)	5 to 30	800 - 1500	Strong water action Mass movement Slight frost action
Savanna	10 to 30	600 - 1250	Running water Moderate wind action
Semi-arid	5 to 30	250 - 600	Strong wind action Running water
Arid	15 to 30	0 - 350	Strong wind action Slight water action

Slopes vary with climate. In general, humid slopes are rounder, due to chemical weathering, whereas arid slopes are jagged or straight owing to mechanical weathering and overland runoff. **Climatic geomorphology** is a branch of geography which studies how different processes operate in different climatic zones and thereby produce different slope forms or shapes.

ASPECT

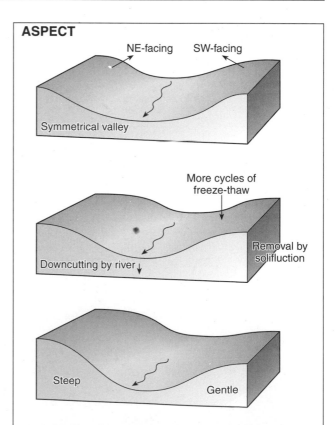

The NE-facing slope remains in the shade. Under periglacial conditions, temperatures rarely rise above freezing. By contrast, the SW-facing slope is subjected to many cycles of freeze-thaw. Solifluction and overland runoff lower the level of the slope, and streams remove the debris from the valley. The result is an asymmetric slope, e.g. the River Exe in Devon, and Clatford Bottom in Wiltshire.

Theories of slope evolution

SLOPE EVOLUTION

Slope evolution refers to the change in slope **form** (shape) over time.

Slopes can be divided into those that are (i) time independent, in which the slope retains a constant angle, although altitude may be lowered, and (ii) time dependent, in which slope angle and altitude decline progressively over time.

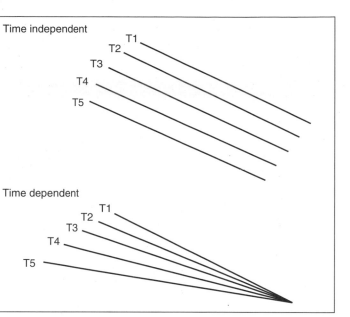

SLOPE DECLINE: W. M. DAVIES

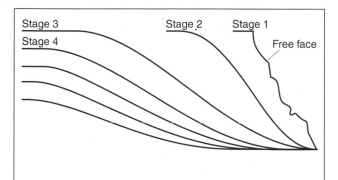

The main processes involved are soil creep, solution, overland runoff, weathering, and fluvial transport at the base. Slopes decline progressively over time. The free face is changed by falls and slumps and develops a regolith. Weathered material is transported by overland runoff and surface wash, eventually producing a convex-concave profile.

SLOPE REPLACEMENT: WALTHER PENCK

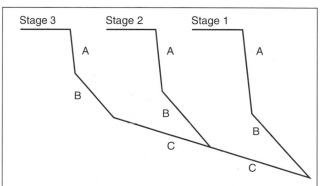

Slope A is replaced by slope B, which is in turn replaced by slope C. Replacement is by lower angle slopes which extend upwards at a constant angle. The segments become increasingly longer as the slope develops. Some free faces may be completely removed. It is common in tectonically active and coastal areas. The size of sediment decreases from A to B to C.

SLOPE RETREAT: L.C. KING

The key elements are a very hard lateritic cap rock controlling the rate of slope evolution. Mechanical weathering and sheet wash are the dominating processes in a semi-arid climate. All elements, except the pediment, retain a constant angle. Pediments vary from 1-2 degrees on fine material to 3-5 degrees on gravel and stony material. It is common in dry areas such as Monument Valley, Arizona.

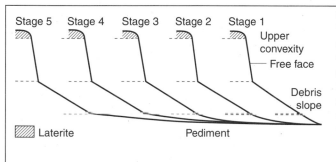

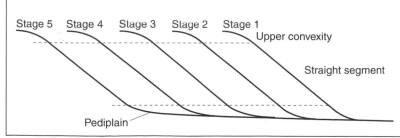

Mass movements

DEFINITION

Mass movements are any large-scale movement of the earth's surface that are not accompanied by a moving agent such as a river, glacier, or ocean wave. They include very small movements, such as soil creep, and fast movements, such as avalanches. They vary from dry movements, such as rock falls, to very fluid movements, such as mudflows.

FLOW

Wet — River

Mudflow

Earthflow

Solifluction

Landslide

Dry

Rockslide — Talus creep — Seasonal soil creep

SLIDE — HEAVE

Fast — Slow

Process	Rate of movement in mm per second
	10^{-7} 10^{-6} 10^{-5} 10^{-4} 10^{-3} 10^{-2} 10^{-1} 10 10^1 10^2 10^3 10^4 10^5
Soil creep	
Solifluction	
Debris flow	
Mudslide	
Flowslide	
Rockfall	

FLOW

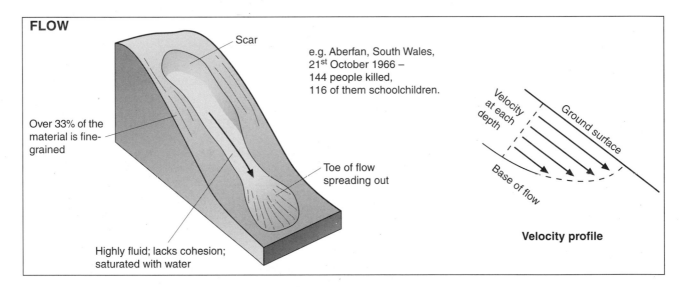

Scar

e.g. Aberfan, South Wales, 21st October 1966 – 144 people killed, 116 of them schoolchildren.

Over 33% of the material is fine-grained

Toe of flow spreading out

Highly fluid; lacks cohesion; saturated with water

Velocity at each depth — Ground surface

Base of flow

Velocity profile

SLIDE

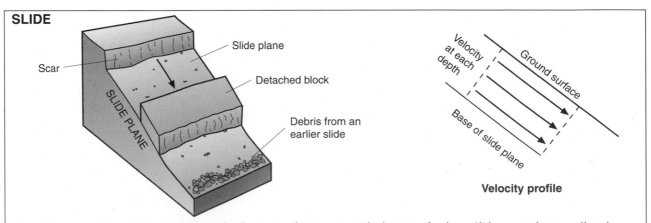

Scar

Slide plane

SLIDE PLANE

Detached block

Debris from an earlier slide

Velocity at each depth — Ground surface

Base of slide plane

Velocity profile

The sliding material maintains its shape and cohesion until it impacts at the bottom of a slope. Slides range from small-scale events close to roads to large-scale movements killing thousands of people, e.g. the Vaiont Dam disaster in Italy where more than 2000 people died on 9th October 1963.

FALLS

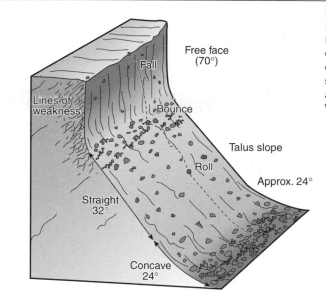

Falls occur on steep slopes (> 70°). The initial cause of the fall may be weathering, e.g. freeze-thaw or disintegration, or erosion prising open lines of weakness. Once the rocks are detached they fall under the influence of gravity. If the fall is short it produces a relatively straight scree; if it is long, it forms a concave scree. A good example of falls and scree is Wastwater in the Lake District.

SLUMPS

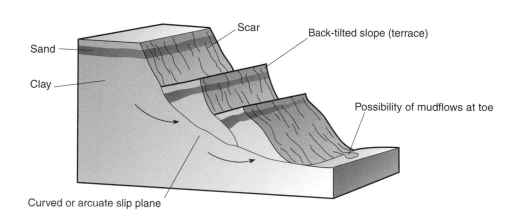

Slumps occur on weaker rocks, especially clay, and have a rotational movement along a curved slip plane. Clay absorbs water, becomes saturated, and exceeds its liquid limit. It then flows along a slip plane. Frequently the base of a cliff has been undercut and weakened by erosion, thereby reducing its strength, e.g. Folkestone Warren. Human activity can also intensify the condition by causing increased pressure on the rocks, e.g. the Holbeck Hall Hotel, Scarborough.

AVALANCHES

Avalanches are rapid movements of snow, ice, rock, or earth down a slope. They are common in mountainous areas: newly-fallen snow may fall off older snow, especially in winter (a **dry avalanche**), while in spring partially-melted snow moves (a **wet avalanche**), often triggered by skiing. Avalanches frequently occur on steep slopes over 22°, especially on north-facing slopes where the lack of sun inhibits the stabilisation of snow. **Debris avalanches** are a rapid mass movement of sediments, often associated with saturated ground conditions.

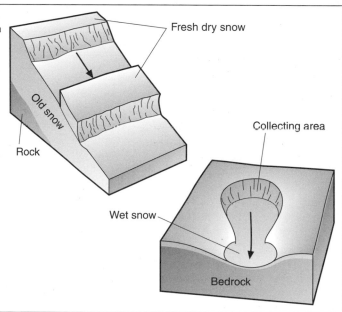

The river basin hydrological cycle

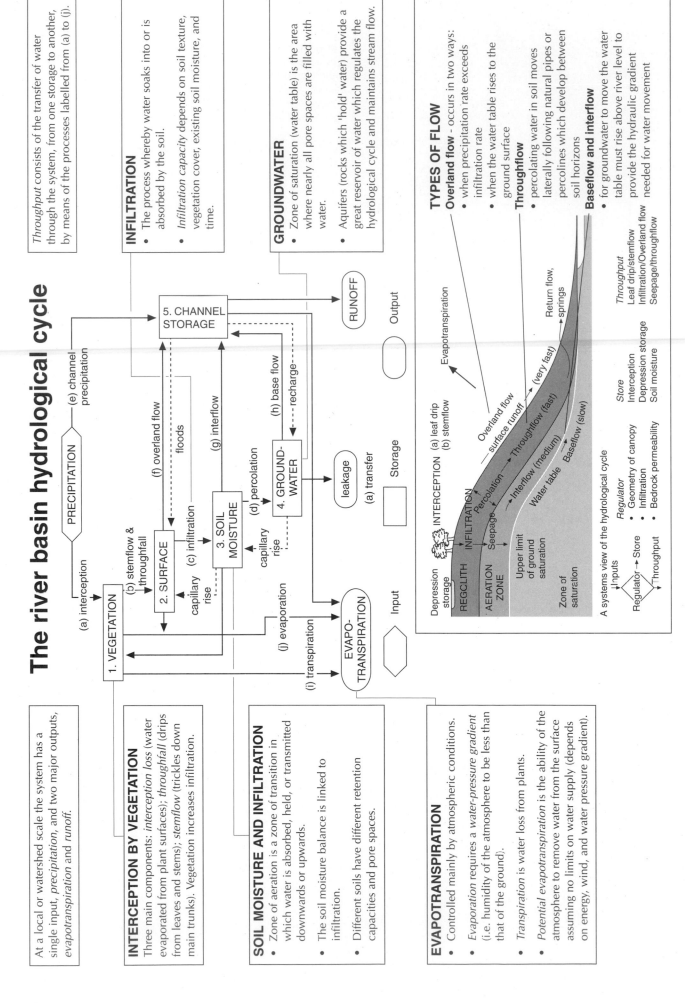

At a local or watershed scale the system has a single input, *precipitation*, and two major outputs, *evapotranspiration* and *runoff*.

Throughput consists of the transfer of water through the system, from one storage to another, by means of the processes labelled from (a) to (j).

INTERCEPTION BY VEGETATION

Three main components: *interception loss* (water evaporated from plant surfaces); *throughfall* (drips from leaves and stems); *stemflow* (trickles down main trunks). Vegetation increases infiltration.

SOIL MOISTURE AND INFILTRATION

- Zone of aeration is a zone of transition in which water is absorbed, held, or transmitted downwards or upwards.
- The soil moisture balance is linked to infiltration.
- Different soils have different retention capacities and pore spaces.

EVAPOTRANSPIRATION

- Controlled mainly by atmospheric conditions.
- *Evaporation* requires a *water-pressure gradient* (i.e. humidity of the atmosphere to be less than that of the ground).
- *Transpiration* is water loss from plants.
- *Potential evapotranspiration* is the ability of the atmosphere to remove water from the surface assuming no limits on water supply (depends on energy, wind, and water pressure gradient).

INFILTRATION

- The process whereby water soaks into or is absorbed by the soil.
- Infiltration capacity depends on soil texture, vegetation cover, existing soil moisture, and time.

GROUNDWATER

- Zone of saturation (water table) is the area where nearly all pore spaces are filled with water.
- Aquifers (rocks which 'hold' water) provide a great reservoir of water which regulates the hydrological cycle and maintains stream flow.

TYPES OF FLOW

Overland flow - occurs in two ways:
- when precipitation rate exceeds infiltration rate
- when the water table rises to the ground surface

Throughflow
- percolating water in soil moves laterally following natural pipes or percolines which develop between soil horizons

Baseflow and interflow
- for groundwater to move the water table must rise above river level to provide the hydraulic gradient needed for water movement

---

PRECIPITATION

(a) interception

(e) channel precipitation

1. VEGETATION

(b) stemflow & throughfall

2. SURFACE

(f) overland flow — floods

(c) infiltration

capillary rise

3. SOIL MOISTURE

(d) percolation

capillary rise

4. GROUND-WATER

(g) interflow

(h) base flow — recharge

5. CHANNEL STORAGE

RUNOFF

leakage

(a) transfer

(j) evaporation

(i) transpiration

EVAPO-TRANSPIRATION

Input

Output

Storage

A systems view of the hydrological cycle

INTERCEPTION — (a) leaf drip, (b) stemflow

Depression storage

REGOLITH

AERATION ZONE

Zone of saturation

Upper limit of ground saturation

INFILTRATION

Percolation

Seepage

Interflow (medium)

Water table

Baseflow (slow)

Overland flow — surface runoff (very fast)

Throughflow (fast)

Return flow, springs

Evapotranspiration

Inputs
Regolith → Store → Throughput

Regulator
- Geometry of canopy
- Infiltration
- Bedrock permeability

Store
Interception
Depression storage
Soil moisture

Throughput
Leaf drip/stemflow
Infiltration/Overland flow
Seepage/throughflow

Drainage basin hydrology

DEFINITION
The drainage basin is the area which drains into a particular river or river system. It is the unit for which a water balance may be constructed to show the disposal of precipitation via interception, soil moisture and groundwater storages, evapotranspiration, and runoff.

BASIN PLAN

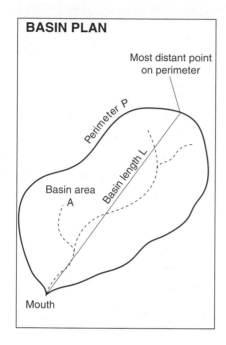

Most distant point on perimeter

Perimeter P

Basin length L

Basin area A

Mouth

DRAINAGE PATTERNS

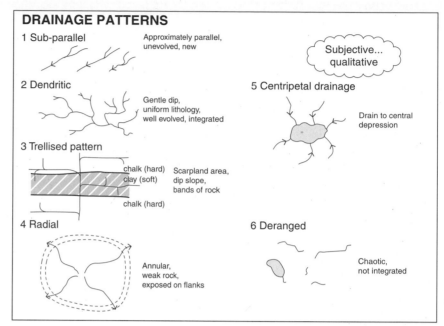

1 Sub-parallel — Approximately parallel, unevolved, new

2 Dendritic — Gentle dip, uniform lithology, well evolved, integrated

3 Trellised pattern — chalk (hard), clay (soft), chalk (hard) — Scarpland area, dip slope, bands of rock

4 Radial — Annular, weak rock, exposed on flanks

5 Centripetal drainage — Drain to central depression

6 Deranged — Chaotic, not integrated

Subjective... qualitative

STREAM ORDERING

Horton analysis
- Smallest finger-tip tributaries are designated order 1; when two first orders join, a channel of order 2 is formed; when two second orders join, a channel of order 3 is formed.

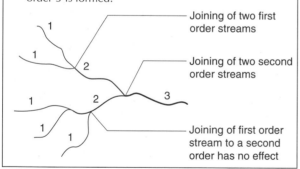

Joining of two first order streams

Joining of two second order streams

Joining of first order stream to a second order has no effect

The Bifurcation Ratio (Rb)
- Law of stream numbers - the higher the order number the lower the number of stream segments.

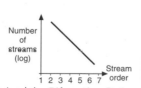

Number of streams (log)

Stream order

1 2 3 4 5 6 7

- From this relationship is derived the Bifurcation Ratio, i.e.

$$Rb - \frac{Nu}{(Nu+1)}$$

Where Rb = bifurcation ration
  Nu = number of streams of a given order
  Nu + 1 = number of streams of the next highest order

DRAINAGE DENSITY (D)

- Drainage density reflects the closeness of spacing of channels and is found by measuring the total lengths of all streams within the basin (ΣL) and dividing by the area of the whole basin (A), i.e. $D = \Sigma L/A$

- It tends to be highest where land surface is impermeable, rainfall is heavy and prolonged, and vegetation cover is lacking.

- Another related measure is *channel frequency* (F) which is defined as the total number of stream segments per unit area.

A

Same drainage density but different channel frequency

B

Same channel frequency but different drainage density

Catchment systems

Different topographical and climatic conditions can lead to varied inputs, throughputs, and processes in the hydrological system.

RELIEF

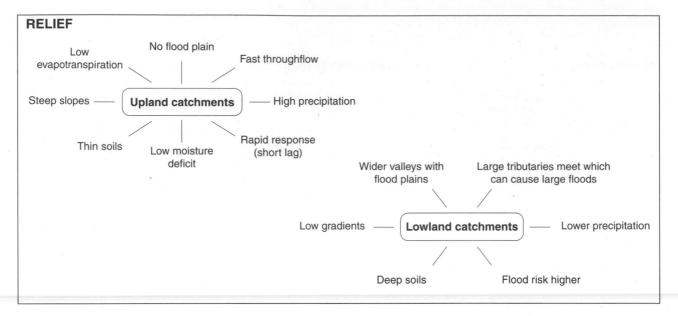

Low evapotranspiration — No flood plain — Fast throughflow

Steep slopes — **Upland catchments** — High precipitation

Thin soils — Low moisture deficit — Rapid response (short lag)

Wider valleys with flood plains — Large tributaries meet which can cause large floods

Low gradients — **Lowland catchments** — Lower precipitation

Deep soils — Flood risk higher

CLIMATE

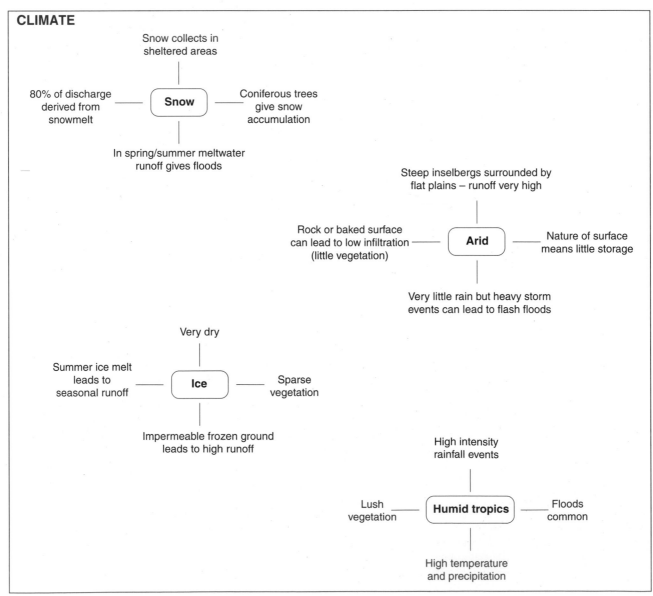

Snow collects in sheltered areas

80% of discharge derived from snowmelt — **Snow** — Coniferous trees give snow accumulation

In spring/summer meltwater runoff gives floods

Steep inselbergs surrounded by flat plains – runoff very high

Rock or baked surface can lead to low infiltration (little vegetation) — **Arid** — Nature of surface means little storage

Very little rain but heavy storm events can lead to flash floods

Very dry

Summer ice melt leads to seasonal runoff — **Ice** — Sparse vegetation

Impermeable frozen ground leads to high runoff

High intensity rainfall events

Lush vegetation — **Humid tropics** — Floods common

High temperature and precipitation

Seasonal variations in river flow: the regime

The pattern of seasonal variation in the flow of a river is known as the *regime*. The regime is related to a number of factors, notably the seasonality of *rainfall*, *temperature*, and *evapotranspiration*. An equatorial river will have a regular regime, but rivers in climates with marked seasonal contrast may have one or more peaks. This can be related to climatic zones.

SIMPLE REGIMES

Simple regimes are where a simple distinction can be made between one period of high water levels and runoff and one period of low water levels and runoff.

Glacier melt
- European mountain rivers have a high-water period (July-August) when glaciers feeding them melt most rapidly.

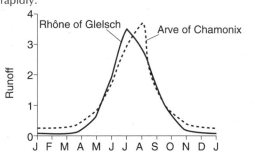

Oceanic rainfall/evapotranspiration
- In many oceanic areas of Europe, rainfall is evenly distributed but high evapotranspiration in summer leads to low runoff.

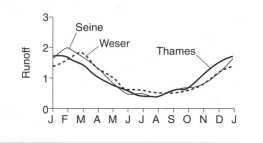

Tropical seasonal rainfall (monsoonal)
- In tropical areas, evapotranspiration tends to be stable (high) but summer rains cause a peak.

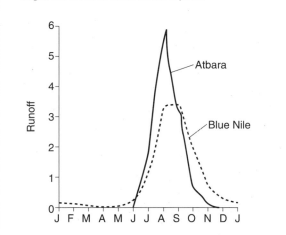

Snowmelt
- Melting of snow cover either in mountainous areas during early summer or over the Great Plains of North America in early/late spring.

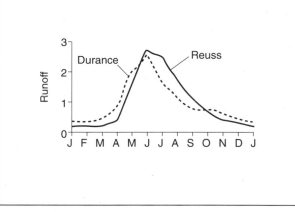

COMPLEX REGIMES

Some rivers are characterised by at least four hydrological phases (two low, two high) which give a more complex regime.

Other rivers, like the Rhine, flow through several distinctive relief regions and receive water from large tributaries which themselves flow over varied terrain. Rivers in this group normally have a single regime in their upper reaches.

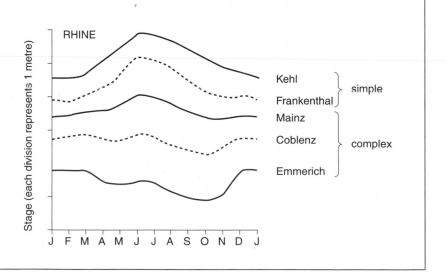

The storm hydrograph

DEFINITIONS

- A *storm hydrograph* measures the speed at which rainfall falling on a drainage basin reaches the river channel. It is a graph on which river discharge during a storm or *runoff* event is plotted against time.

- *Discharge* (Q) is the volume of flow passing through a cross-section of a river during a given period of time (usually measured in cumecs - m³/s).

READING A STORM HYDROGRAPH

Discharge peak

- higher in larger basins

- 'steep catchments' will have lower infiltration rates so high peaks

- 'flat catchments' will have high infiltration so more throughflow and lower peaks

Hydrograph size (area under the graph)

- the higher the rainfall, the greater the discharge

- the larger the basin size, the greater the discharge

Recession limb

- indicates the amount of groundwater depletion caused by throughflow

- influenced by geological composition and behaviour of local aquifers

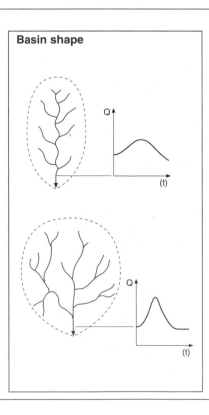

Lag time

- time interval between peak rainfall and peak discharge

- influenced by basin shape, steepness, stream order

- relationship between runoff and throughflow is main determinant

Runoff

- reveals the relationship between overland flow and throughflow

- where infiltration is low and rainfall strong overland flow will dominate

Baseflow

- the seepage of groundwater into the channel

- a slow movement which is the main, long-term supplier of the river's discharge

INFLUENCES

Climate

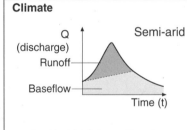

- short but high intensity rain
- low infiltration
- overland flow is dominant

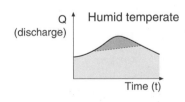

- deep soils and abundant vegetation
- high infiltration
- throughflow is dominant

Basin shape

Land use

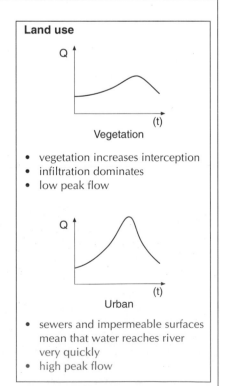

- vegetation increases interception
- infiltration dominates
- low peak flow

- sewers and impermeable surfaces mean that water reaches river very quickly
- high peak flow

Urban hydrology

Storm-water sewers
- reduce the distance storm-water must travel before reaching a channel
- increase the velocity of flow because sewers are smoother than natural channels
- reduce storage because sewers are designed to drain quickly

THE EFFECTS OF URBANISATION ON HYDROLOGICAL PROCESSES

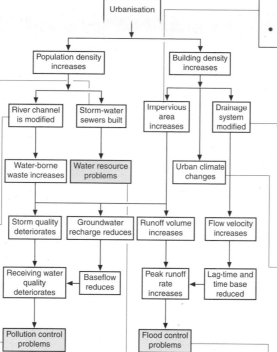

Urbanisation

Population density increases → Building density increases

River channel is modified | Storm-water sewers built | Impervious area increases | Drainage system modified

Water-borne waste increases | Water resource problems | Urban climate changes

Storm quality deteriorates | Groundwater recharge reduces | Runoff volume increases | Flow velocity increases

Receiving water quality deteriorates | Baseflow reduces | Peak runoff rate increases | Lag-time and time base reduced

Pollution control problems | Flood control problems

Replacement of vegetated soils with impermeable surfaces
- reduces storage and so increases runoff
- increases velocity of overland flow
- decreases evapotranspiration because urban surfaces are usually dry
- reduces infiltration and percolation

Building activity
- clears vegetation which exposes soil and increases overland flow
- disturbs and dumps the soil, increasing erodibility
- eventually protects the soil with armour of concrete or tarmac

Rainfall climatology of urban areas
- greater aerodynamic roughness and urban heat island
- more rainfall, especially in summer
- heavier and more frequent thunderstorms

Encroachment on the river channel
- embankments, reclamation, and river-side roads
- usually reduces channel width leading to higher floods
- bridges can restrict free discharge of floods and increase levels upstream

Pollution control problems
- storm-water that washes off roads and roofs can contain heavy metals, volatile solids, and organic chemicals
- annual runoff from 1 km of the M1 included 1.5 tons of suspended sediment, 4 kg of lead, 126 kg of oil, and 18 kg of aromatic hydrocarbons

Water resource problems
- groundwater recharge may be reduced because sewers bypass the mechanisms of percolation and seepage
- groundwater abstraction through wells may also reduce the store locally
- irrigation can draw on water resources leading not only to depletion but also pollution

Flood control problems
- urbanisation increases the peak of the mean annual flood, especially in moderate conditions
- a 243 per cent increase resulted from the building of Stevenage New Town
- however, during heavy prolonged rainfall, saturated soil behaves in a similar way to urban surfaces

URBAN HYDROLOGY AND THE STORM HYDROGRAPH

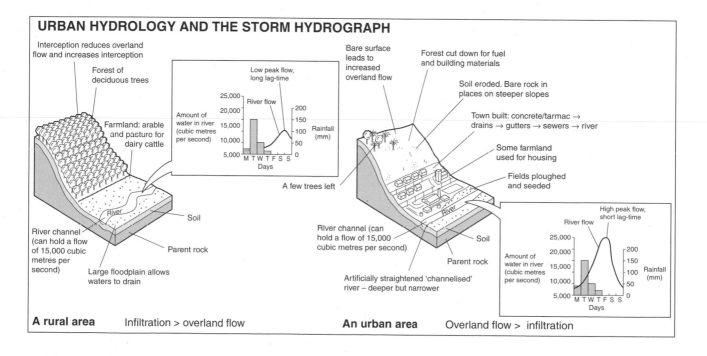

Interception reduces overland flow and increases interception

Forest of deciduous trees

Low peak flow, long lag-time

River flow

Amount of water in river (cubic metres per second)

Rainfall (mm)

Days

Farmland: arable and pasture for dairy cattle

River channel (can hold a flow of 15,000 cubic metres per second)

Large floodplain allows waters to drain

Soil

Parent rock

A rural area   Infiltration > overland flow

Bare surface leads to increased overland flow

Forest cut down for fuel and building materials

Soil eroded. Bare rock in places on steeper slopes

Town built: concrete/tarmac → drains → gutters → sewers → river

Some farmland used for housing

Fields ploughed and seeded

A few trees left

River channel (can hold a flow of 15,000 cubic metres per second)

Artificially straightened 'channelised' river – deeper but narrower

High peak flow, short lag-time

River flow

Amount of water in river (cubic metres per second)

Rainfall (mm)

Days

Soil

Parent rock

An urban area   Overland flow > infiltration

The long profile

By plotting a line graph of river's height above base level against distance from its source the long profile is revealed. As rivers evolve through time and over distance the stream passes through a series of distinct stages.

Key
– processes
– *landforms*

Upper course
vertical erosion
weathering
headward erosion

- V-shaped valley
- pot-holes
- interlocking spurs
- waterfalls
- rapids
- gorges

Middle course
lateral erosion
transportation

- asymmetrical channel
- floodplain
- truncated spurs
- meanders
- river cliff/ slip-off slope

Lower course
transportation
deposition

- large channel
- braiding
- ox-bow lakes
- large floodplain
- bluffs
- levels
- deltas

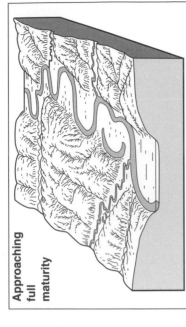

Youth

A In the initial stage a stream has lakes, waterfalls, and rapids.
- initial uplift
- overland flow is concentrated in depressions making lakes
- overflow links lake basins and initial stream system forms

Middle youth

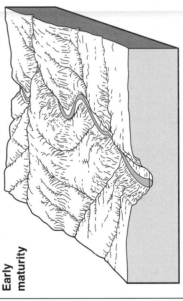

B By middle youth the lakes are gone, but falls and rapids persist along the narrow incised gorge.
- deepening of channel
- waterfalls evolve into rapids
- headward erosion can form gorges
- exogenetic load through landslides and weathering

Early maturity

C Early maturity brings a smoothly graduated profile without rapids or falls, but with the beginnings of a floodplain.
- smooth, even gradient
- floodplain begins to form

Approaching full maturity

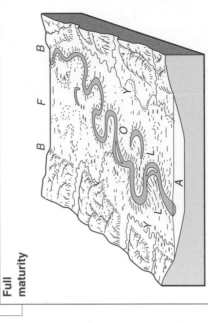

D Approaching full maturity, the stream has a floodplain almost wide enough to accomodate its meanders.
- river is in equilibrium
- supply of load = average rate at which stream can transport load
- floodplain gets wider through the enlargement and downstream migration of meanders

Full maturity

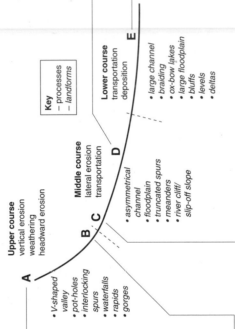

E Full maturity is marked by a broad floodplain and freely developed meanders. *L = levee; O = ox-bow lakes; Y = yazoo stream; A = alluvium; B = bluffs; F = floodplain.*

Rivers as sediment systems

The river is a sub-system of a large unit - the drainage basin. The sediment system within a river is a further sub-system depending on many variables:

- discharge
- climate
- relief
- rock type

Sources

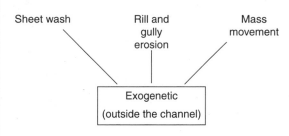

Sheet wash   Rill and gully erosion   Mass movement

Exogenetic
(outside the channel)

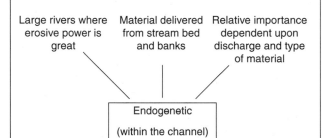

Large rivers where erosive power is great   Material delivered from stream bed and banks   Relative importance dependent upon discharge and type of material

Endogenetic
(within the channel)

In upland areas discharge is low and bed material is large, so endogentic sediment yield is low.

In lowland areas discharge is high and bed and bank material is not resistant, so sediment yield will be high.

CRITICAL EROSION VELOCITY

The critical erosion velocity curve is the range of velocity needed to pick up particles of various sizes. The relationship between velocity and particle erosion, transportation, and deposition is given by the Hjulstrom Curve.

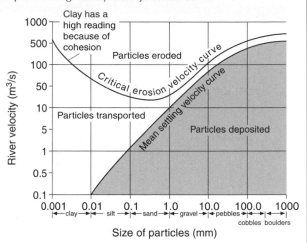

EQUILIBRIUM

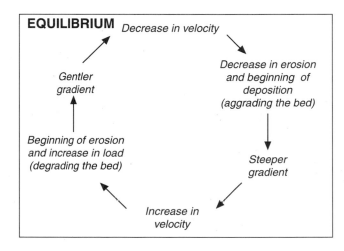

Decrease in velocity

Decrease in erosion and beginning of deposition (aggrading the bed)

Steeper gradient

Increase in velocity

Beginning of erosion and increase in load (degrading the bed)

Gentler gradient

THE RIVER'S LOAD

Suspended sediment load
- carried with the body of the current
- 'wash-load' - small silt clay particles (<0.0625 mm)
- 'suspended bed material' - larger fine-medium sands which derive from channel bed

Bedload
- moves by sliding, rolling, or saltating
- maybe exogenetic or endogenetic

Dissolved load
- derives from precipitation, chemical weathering, erosion, atmospheric fallout, mineral springs, pollution

Transportation

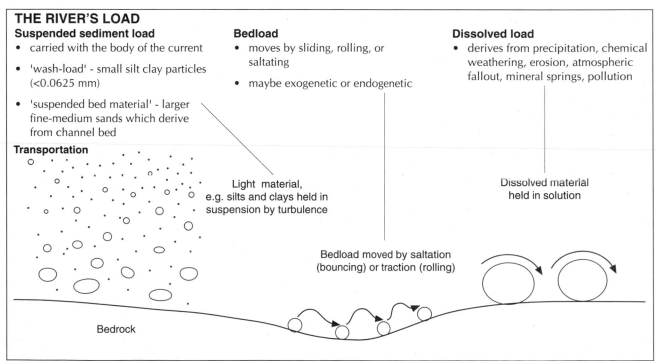

Light material, e.g. silts and clays held in suspension by turbulence

Dissolved material held in solution

Bedload moved by saltation (bouncing) or traction (rolling)

Bedrock

The river channel

TYPES OF FLOW

Streamflow is very complex. The velocity and therefore the energy is controlled by:

- *gradient* of channel bed
- *volume of water* within the channel
- the *shape* of the channel
- *channel roughness*/friction

LAMINAR FLOW

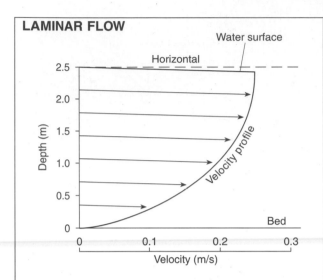

Necessary conditions for laminar flow include:

- smooth, straight channel
- shallow water
- low, non-uniform velocity allowing water to flow in sheets parallel to channel bed. Rare in reality. Most common in lower reaches

TURBULENT FLOW

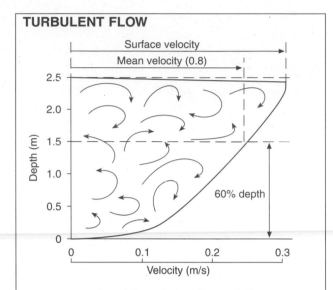

Necessary conditions for turbulent flow include:

- complex channel shape, e.g. winding channels, riffles and pools
- high velocity turbulence is associated with *cavitation* as *eddies* trap air in pores, cracks, and crevices which is then released under great pressure

CHANNEL ROUGHNESS

Channel roughness causes friction which slows the flow of the river. Friction is caused by boulders, vegetation, sinuosity, and bedform. It is measured using *Manning's 'n'* which expresses the relationship between channel roughness and velocity in an equation:

$$v = \frac{R^{\frac{2}{3}} S^{\frac{1}{2}}}{'n'}$$

v = velocity
R = hydraulic radius
S = slope
n = roughness

The higher the value the rougher the bed, e.g.

Bed profile	Sand and gravel	Coarse gravel	Boulders
Uniform	0.02	0.03	0.05
Undulating	0.05	0.06	0.07
Irregular	0.08	0.09	0.10

CHANNEL SHAPE

Channel efficiency is measured by the *hydraulic radius*, i.e. *cross-sectional area*. This is wetted perimeter and is affected by river level and channel shape.

River level

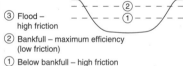

③ Flood – high friction
② Bankfull – maximum efficiency (low friction)
① Below bankfull – high friction

Shape

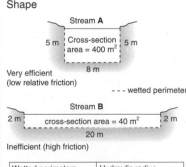

Stream **A**
5 m | Cross-section area = 400 m² | 5 m
8 m
Very efficient (low relative friction)

- - - wetted perimeter

Stream **B**
2 m | cross-section area = 40 m² | 2 m
20 m
Inefficient (high friction)

Wetted perimeters	Hydraulic radius
Stream **A**: 5 + 5 + 8 = 18 m	Stream **A**: $\frac{40}{18}$ = 2.22 m
Stream **B**: 2 + 2 + 20 = 24 m	Stream **B**: $\frac{40}{24}$ = 1.66 m

CHANNEL SLOPE

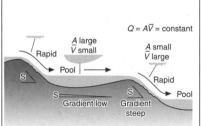

Discharge (Q) - the volume of water passing through a given cross-section in a given unit of time; velocity (v); cross-sectional area (A).

Steeper gradients should lead to higher velocities because of gravity. Velocity increases quickly where a stream passes from a pool of low gradient to a steep stretch of rapids.

As v increases cross-sectional area decreases; as v decreases cross-sectional area increases.

Meanders

WHAT CAUSES MEANDERS?

There is no single explanation, but a number of factors have been suggested:

- **sandbars** - flume tank experimentation suggests that sinuosity can be triggered by sandbars

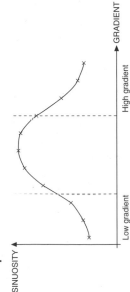

Channel shifts

Sandbar

- **slope thresholds**

SINUOSITY

High gradient

Low gradient

GRADIENT

- **helicoidal flow**

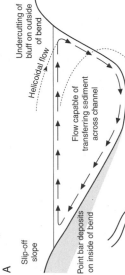

A

B

Slip-off slope

Undercutting of bluff on outside of bend

Helicoidal flow

Flow capable of transferring sediment across channel

As well as being transverse, the flow must continue downstream. Flow therefore is helicoidal

Point bar deposits on inside of bend

A — B

The thalweg is the line tracing the deepest and fastest water. The thalweg moves from side to side within the channel, and also corkscrews in cross-section. This is helicoidal flow and increases the amplitude of the meander.

SINUOSITY

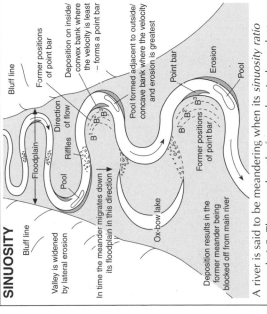

Bluff line

Valley is widened by lateral erosion

Floodplain

Bluff line

Former positions of point bar

Deposition on inside/convex bank where the velocity is least — forms a point bar

Direction of flow

Riffles

Pool

In time the meander migrates down its floodplain in this direction

Pool formed adjacent to outside/concave bank where the velocity and erosion is greatest

Point bar

B¹ B² B³

B¹ B² B³

Former positions of point bar

Ox-bow lake

Erosion

Pool

Deposition results in the former meander being blocked off from main river

A river is said to be meandering when its *sinuosity ratio* exceeds 1.5. The *wavelength* of meanders is dependent on three major factors: channel width, discharge, and the nature of the bed and banks.

POOLS AND RIFFLES

Stage 1 → Stage 2 → Stage 3 → Stage 4 → Stage 5

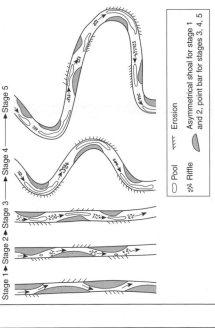

- Pool
- Riffle ⚏ Erosion
- Asymmetrical shoal for stage 1 and 2, point bar for stages 3, 4, 5

Meanders have strong links with pools and riffles. They are caused by turbulence. Roller eddies cause deposition of coarse sediment (riffles) at high velocity points and fine sediment (pools) at points of low velocity. Riffles have a steeper gradient than pools which leads to sinuosity.

IN WHAT WAYS DO MEANDERS MIGRATE?

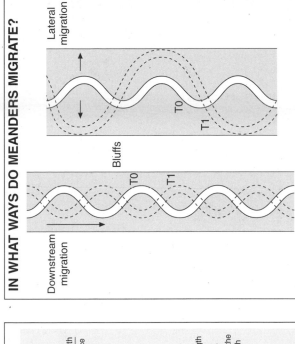

Lateral migration

T0

T1

Bluffs

Downstream migration

T0

T1

This results in a greatly widened *floodplain*.

Development of a meander through time

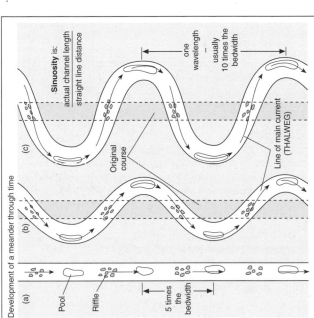

(a)

Pool

Riffle

5 times the bedwidth

(b)

(c)

Original course

Sinuosity is: $\dfrac{\text{actual channel length}}{\text{straight line distance}}$

one wavelength — usually 10 times the bedwidth

Line of main current (THALWEG)

Deltas and estuaries

DELTAS

Deltas are formed when river sediments are deposited when a river enters a standing body of water such as a lake, lagoon, or ocean. Deposition occurs because velocity is checked (see Hjulstrom Curve). A number of factors affect the formation of deltas:

- amount and calibre (size) of load - rivers must be heavily laden, coarse sediment will be dumped first

- salinity - salt-water causes charged particles in freshwater to flocculate or adhere together

- gradient of coastline - delta formation will be more likely on gentle coastline where turbidity is less

- vegetation - plant life will slow waters and so increase deposition, and also provide a surface on which deposition can occur

- low energy river discharge and/or low energy wave or tidal energy

Structure of a simple delta

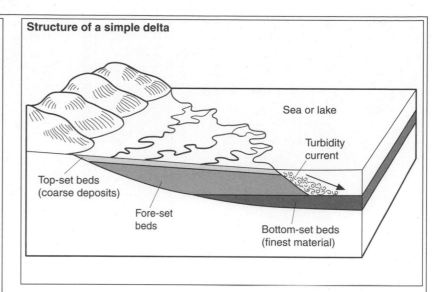

Deltas occur in three forms:
(i) arcuate - many distributaries which branch out radially

(ii) cuspate - a pointed delta formed by a dominant channel

(iii) bird's foot - long, projecting fingers which grow at the ends of distributaries

ESTUARIES

Estuaries occur where a coastal area has recently subsided or the ocean level has risen, causing the lower part of the river to be drowned. Unlike a *ria*, which is also a drowned river valley, estuaries form traps for sediment which may be exposed some or all of the time.

Vegetation

- vegetation like *Spartina towsendii* stabilises the loose surface of the sediment preventing erosion

- it also reduces velocity and so increases deposition

Landforms, sediments, and water movements in estuaries

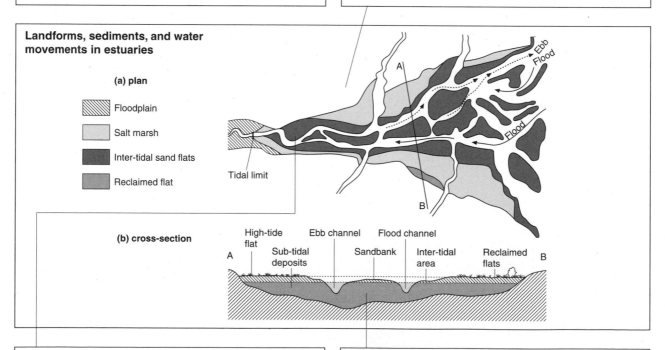

Tidal range

- rising *flood* tides and falling *ebb* tides form channels

- velocity of water flow may be great in both directions

- the result is a shifting set of channels based on erosion and deposition

Sedimentation

- estuaries are sheltered which enhances sedimentation due to reduced velocity

- freshwater meeting salt-water may lead to *flocculation* where charged particles cluster and sink

Rivers and people

CAUSES OF RIVER 'PROBLEMS'

INHERENT   River channel changes, e.g. migration
               cut-offs

EXTERNAL   Climatic change, anthropogenic causes

NATURAL   Storm and flood occurrence -
               frequency, timing, magnitude
               Climatic change

HUMAN ACTIVITIES
    Direct   River regulation
               Channelisation
               Water abstraction
               Waste disposal
               Irrigation
               Drainage, especially agricultural
               Dams

    Indirect   Land-use change,
               especially deforestation,
               afforestation
               Urbanisation and roads
               Mining
               Agricultural practices

PROBLEMS RELATED TO RIVERS

Problem	Effects
FLOODING	→ Destruction of structures and communications Danger to life and property Destruction of crops Interruption of activities Drainage difficulties
EROSION	→ Destruction of structures Loss of land and property Boundary disputes
SEDIMENTATION	→ Flooding Drainage pattern alteration Change in ecology Water quality change Navigation difficulties
CHANNEL AND CHANGES INSTABILITY	→ Destruction of structures Loss of amenity value Navigation difficulties Boundary disputes
ECOLOGICAL CHANGES	→ Loss of amenity value Decrease in fish stocks
LAND DRAINAGE	→ Groundwater alterations Vegetation change Agricultural change

CASE STUDY: THE THREE GORGES DAM

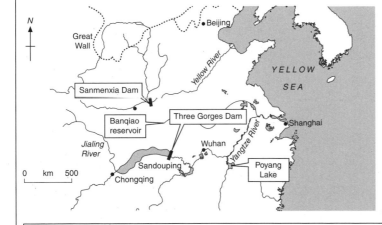

The decision to build the Three Gorges Dam on the Yangtze River highlights some of the conflicts apparent in the way people use the river. The dam will enable China to:

- generate up to 18,000 megawatts of power, reducing the country's dependence on coal

- supply Shanghai's 13 million people with water

- protect 10 million people from flooding (over 300,000 people have died in China as a result of flooding this century)

- water levels will be raised to allow shipping above the Three Gorges (formally rapids)

Protest against the Three Gorges Dam

- Most floods in recent years have come from rivers which join the Yangtze below the Three Gorges Dam.

- The region is seismically active and landslides are frequent.

- The port at the head of the lake may become silted up as a result of increased deposition and the development of a delta at the head of the lake.

- Up to 1.2 million people will have to be moved to make way for the dam.

- Much of the land available for resettlement is over 800 m above sea-level, and is colder with infertile thin soils on relatively steep slopes.

- Dozens of towns, for example Wanxian and Fuling with 140,000 and 80,000 people respectively, will be flooded.

- Up to 530 million tonnes of silt are carried through the Gorge annually - the first dam on the river lost its capacity within seven years and one on the Yellow River filled with silt within four years.

- To reduce the silt load afforestation is needed, but resettlement of people will cause greater pressure on the slopes above the dam.

- The dam will interfere with aquatic life - the Siberian Crane and the White Flag Dolphin are threatened with extinction.

- Archaeological treasures will be drowned, including the Zhang Fei temple.

Sand dunes in hot deserts

Sand-sized particles (0.15-2.0 mm) are moved by three processes:

- **suspension** (<0.15 mm) - particles light enough to be carried substantial distances by the wind

- **saltation** (0.15-0.25 mm) - a rolling particle gains sufficient velocity for it to leave the sand surface in one or more 'jumps'

- **surface creep** (0.25-2.0 mm) - larger grains are dislodged by saltating grains

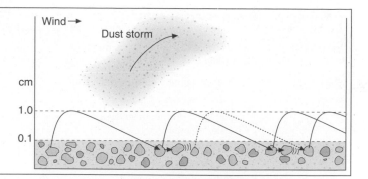

TYPES OF DUNE

Dune morphology depends on wind direction and speed, the supply of sand, availability of vegetation, and the nature of the ground surface.

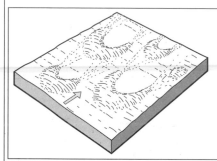

1 Barchan dunes

Broadly rounded ends point downwind. Windward side is gentle, being the slope up which the limited supply of sand moves.

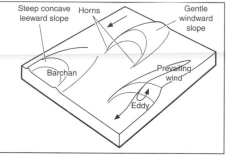

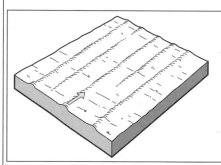

2 Parabolic dunes

Owe presence to limited vegetation or soil moisture. Hairpin-shaped with nose pointing downwind. They are mobile and may have saucer shaped 'blow-outs' caused by deflation.

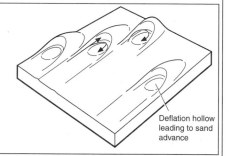

3 Longitudinal or Seif dunes

Limited sand but strong winds with a seasonal change in direction form dune ridges 5-30 m high and may extend for tens or hundreds of kilometres. Vortex flow has been suggested as mechanism for their formation.

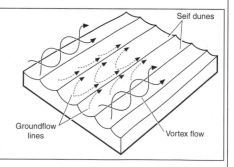

4 Obstacle or topograhpic dunes

(a) Sand dune in shadow of obstacle

(b) Sand drift in lee of gap in obstacle

(c) Wrap-around dune

(d) Falling, climbing, and echo dunes

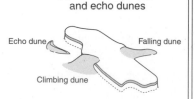

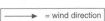
= wind direction

Distribution of arid and semi-arid environments

In desert areas a shortage of rainfall and high temperatures lead to a *soil water deficit*. Aridity is defined using the water balance. In arid regions there is a deficit in *water balance* over the year. In semi-arid areas the water balance fluctuates between the positive and negative.

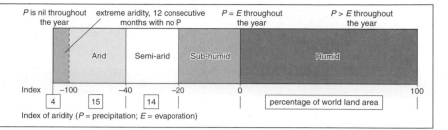

P is nil throughout the year	extreme aridity, 12 consecutive months with no P		$P = E$ throughout the year	$P > E$ throughout the year
	Arid	Semi-arid	Sub-humid	Humid

Index   −100   −40   −20   0   100

| 4 | 15 | 14 | | percentage of world land area |

Index of aridity (P = precipitation; E = evaporation)

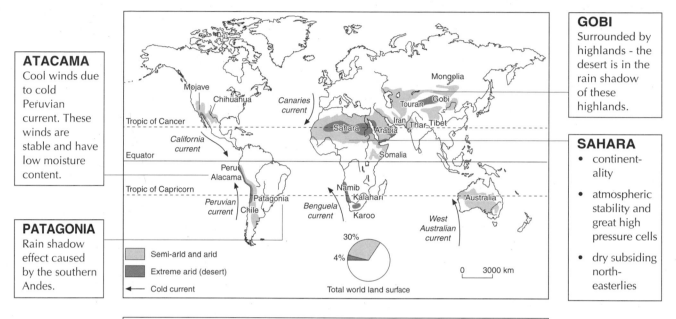

ATACAMA
Cool winds due to cold Peruvian current. These winds are stable and have low moisture content.

PATAGONIA
Rain shadow effect caused by the southern Andes.

GOBI
Surrounded by highlands - the desert is in the rain shadow of these highlands.

SAHARA
- continentality
- atmospheric stability and great high pressure cells
- dry subsiding north-easterlies

Map labels: Mojave, Chihuahua, Canaries current, Mongolia, Touran, Gobi, Tropic of Cancer, California current, Iran, Thar, Tibet, Sahara, Arabia, Equator, Somalia, Peru, Alacama, Tropic of Capricorn, Peruvian current, Chile, Patagonia, Namib, Kalahari, Benguela current, Karoo, Australia, West Australian current

Semi-arid and arid
Extreme arid (desert)
← Cold current

30%
4%
Total world land surface

0   3000 km

Causes of aridity - continentality, the rain shadow effect, and cold ocean currents all promote aridity.

A SEMI-ARID ENVIRONMENT - THE SAHEL

If north-easterlies dominate the result is drought. If moist south-westerlies push up from the south the result is seasonal rain.

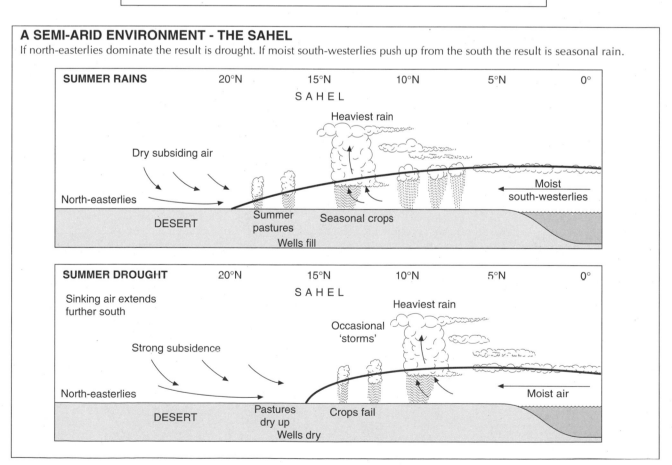

SUMMER RAINS   20°N   15°N   10°N   5°N   0°

S A H E L

Heaviest rain

Dry subsiding air

North-easterlies

Moist south-westerlies

DESERT   Summer pastures   Seasonal crops

Wells fill

SUMMER DROUGHT   20°N   15°N   10°N   5°N   0°

S A H E L

Sinking air extends further south

Heaviest rain

Occasional 'storms'

Strong subsidence

North-easterlies

Moist air

DESERT   Pastures dry up   Crops fail

Wells dry

Landforms of the hot desert

PEDIMENTS

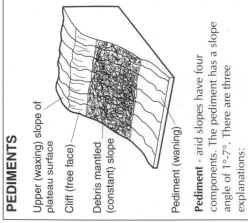

- Upper (waxing) slope of plateau surface
- Cliff (free face)
- Debris mantled (constant) slope
- Pediment (waning)

Pediment - arid slopes have four components. The pediment has a slope angle of 1°-7°. There are three explanations:

- sheets of water at high velocity planed down the surface
- lateral planation where streams flowing from the mountains swing from side to side eroding the surface
- surface and sub-surface weathering

OASES

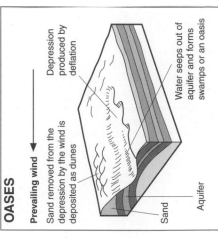

- Prevailing wind ▶
- Depression produced by deflation
- Sand removed from the depression by the wind is deposited as dunes
- Water seeps out of aquifer and forms swamps or an oasis
- Sand
- Aquifer

Oasis - some hollows produced by deflation reach down to water-bearing rocks forming an oasis.

WADIS

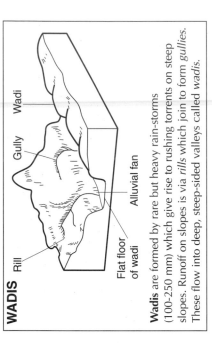

- Wadi
- Gully
- Alluvial fan
- Rill
- Flat floor of wadi

Wadis are formed by rare but heavy rain-storms (100–250 mm) which give rise to rushing torrents on steep slopes. Runoff on slopes is via *rills* which join to form *gullies*. These flow into deep, steep-sided valleys called *wadis*.

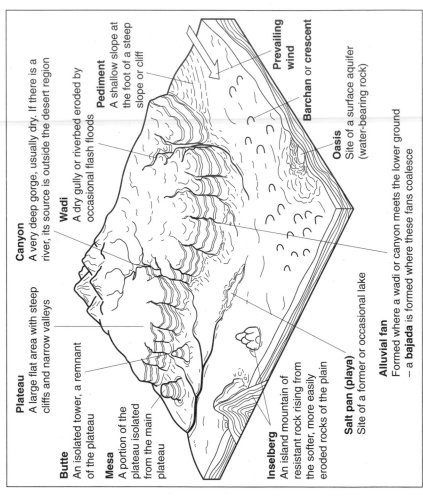

Plateau
A large flat area with steep cliffs and narrow valleys

Butte
An isolated tower, a remnant of the plateau

Mesa
A portion of the plateau isolated from the main plateau

Canyon
A very deep gorge, usually dry. If there is a river, its source is outside the desert region

Wadi
A dry gully or riverbed eroded by occasional flash floods

Pediment
A shallow slope at the foot of a steep slope or cliff

Prevailing wind

Barchan or crescent

Oasis
Site of a surface aquifer (water-bearing rock)

Inselberg
An island mountain of resistant rock rising from the softer, more easily eroded rocks of the plain

Salt pan (playa)
Site of a former or occasional lake

Alluvial fan
Formed where a wadi or canyon meets the lower ground – a **bajada** is formed where these fans coalesce

INSELBERGS

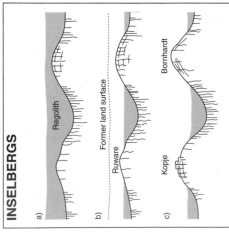

- a) Regolith
- Former land surface
- b) Ruware
- Bornhardt
- c) Kopje

Inselbergs can be formed by gradual slope retreat (backwearing or *pediplanation*) or by *exhumation* whereby surface weathering occurs in conjunction with deep weathering. The *basal surface* is exposed and further surface stripping leads to steep-sided hills.

ALLUVIAL FANS

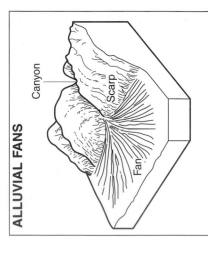

- Canyon
- Scarp
- Fan

Alluvial fans are cones of sediment which form due to a loss of energy as the river emerges from a constrained channel.

Glaciation

GLACIAL SYSTEMS

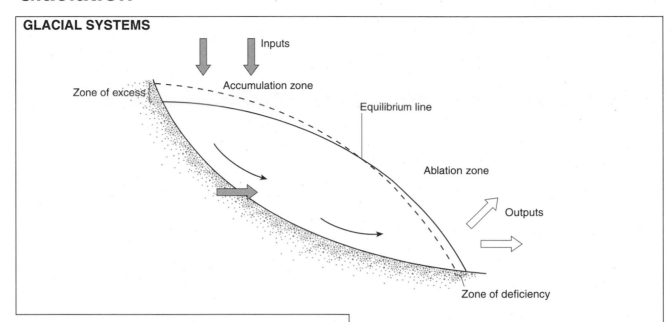

HOW SNOW BECOMES ICE

Initially snow falls as flakes and the accumulation is known as **alimentation**. With continued alimentation the lower snowflakes are compressed under more snow and gradually change into a collection/aggregate of granular ice pellets called **neve** or firn. Increased pressure causes the flakes to melt. Meltwater seeps into the gaps between the pellets and freezes. Some places still have air, which gives the neve a white colour. With continued **accumulation** and pressure the neve becomes tightly packed, and any remaining air is expelled, its place filled with freezing water. Some molecules of water vapour may condense straight into ice, again filling any spaces. Thus, the neve changes into glacier ice, by **compaction** and **crystallisation**, with a characteristic bluish colour. The change from neve to ice occurs, typically, when the neve is 30 m thick. Glacier ice in bulk is a 'granular aggregate of interlocking grains', each grain being an ice crystal. Between each crystal there remains an extremely thin 'intergranular film', consisting of a water-like solution containing chlorides and other salts. The presence of these salts lowers the freezing point around the crystals and so keeps the solution in a liquid state. This film thus acts as a lubricant, aiding ice movement.

A glacial system is the balance between inputs, storage, and outputs. Inputs include **accumulation** of snow, avalanches, debris, heat, and meltwater. The main store is that of ice, but the glacier also carries debris, called **moraine**, and meltwater. The outputs are the losses due to **ablation**, the melting of snow and ice, and sublimation of ice to vapour, as well as sediment.

The **regime** of the glacier refers to whether the glacier is advancing or retreating:

if accumulation > ablation, the glacier advances
if accumulation < ablation, the glacier retreats
if accumulation = ablation, the glacier is steady

Glacial systems can be studied on an annual basis or on a much longer time scale. The size of a glacier depends on its regime, i.e. the balance between the rate and amount of supply of ice and the amount and rate of ice loss. The glacier will have a **positive regime** when the supply is greater than loss by ablation (melting, evaporation, calving, wind erosion, avalanche, and so on) and so the glacier will thicken and advance. A **negative regime** will occur when the wasting is greater than the supply (e.g. the Rhone glacier today) and thus the glacier will thin and retreat. Any glacier can be divided into two sections: an area of accumulation at high altitudes generally, and an area of ablation at the snout.

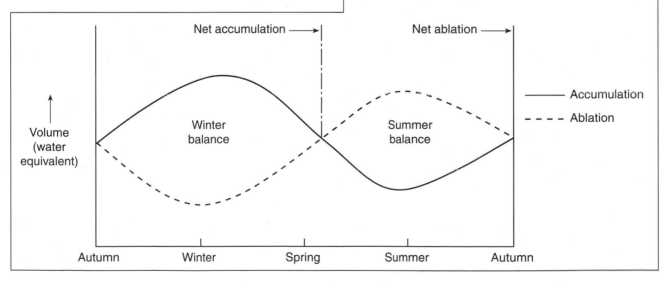

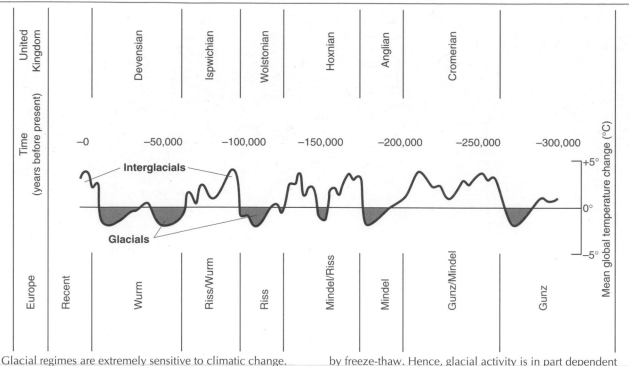

Glacial regimes are extremely sensitive to climatic change. Over the last two million years - the **Quaternary Era** - there have been cold **glacial** phases in which glaciers were present or advancing, and warm **interglacial** periods when ice retreated. This is important in explaining the effectiveness of glacial erosion and transport. Each glacial period is preceded and followed by a **periglacial** period, characterised by intense freeze-thaw activity, nivation, and snowmelt. This breaks up the landscape and the glacier is then able to erode and transport the prepared material. Moreover, at the end of each glacial period, there is pressure release, or dilation, whereby the underlying rocks expand and break as the weight of the glacier is removed. This exposes an even greater area of bedrock to be attacked by freeze-thaw. Hence, glacial activity is in part dependent upon processes that operate in the periglacial stages.

CAUSES OF GLACIAL PERIODS AND ICE ADVANCES

Many different causes of these climatic changes have been proposed and include changes in the earth's orbit, the tilt of the earth's axis, solar radiation, the position of the poles, the amount of water vapour in the atmosphere, the distribution of land and sea, the relative levels of land and sea, the nature and direction of ocean currents, and a reduction of CO_2 in the atmosphere.

Studying the evidence for glacial activity is complicated for a number of reasons: there have been many glaciations; the later glaciations remove the evidence of earlier ones; it is not possible or practical to study processes happening under glaciers; and many of the landforms have been subsequently modified by mass movements and fluvial activity.

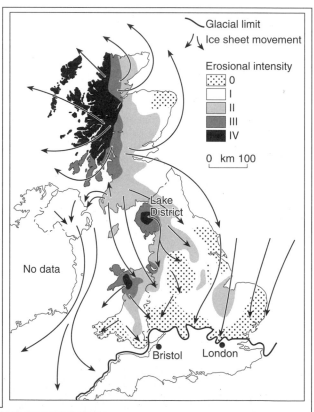

PAST AND PRESENT DISTRIBUTION OF GLACIAL ENVIRONMENTS

It is possible to piece together the main actions of glacial activities during the **Pleistocene Ice Age** (2 m.y. b.p.). Most of Britain, Ireland, the North Sea, Scandinavia, and northern Europe was covered by an ice sheet, whereas the advance in Alpine areas was more limited. During glacial advances the summer temperature remained below 0°C allowing snow and ice to remain all year. In Britain, glaciers pushed as far south as the Bristol Channel. Upland areas were affected by more glacial periods and more intense activity. The main advances include the Anglian Glaciation between c.425,000 and 380,000 years ago, the Wolstonian Glaciation 175,000 to 128,000 years ago, and the Devensian advances between 26,000 and 15,000 years ago and between 12,000 and 10,000 years ago. The latter, the Loch Lomond readvance, was limited to western Scotland. Although there is much debate regarding the precise timings of these phases, and in some cases even the sequence, the effect of **multiple glaciations** on the environment is clear.

Thermal classification and glacier movement

CLASSIFICATION

i) **Temperate glaciers** are warm-based glaciers. Water is present throughout the ice mass and acts as a lubricant allowing ice to move freely and erode the rock. The heat is a result of the pressure of the ice or meltwater percolating down through the ice and/or release from the underlying bedrock.

ii) **In polar or cold-based glaciers** the ice remains frozen at the base, and consequently there is little water or movement. Very little erosion results.

Temperate glaciers generally have velocities of between 20 m and 200 m per year, but can reach speeds of up to 1000 m/yr. By contrast, polar glaciers may advance at only a few metres per year.

GLACIER VELOCITY

The velocity of a glacier is controlled by (i) the **gradient** of the rock floor, (ii) the **thickness of the ice** (which controls pressure and meltwater), and (iii) **temperature** within the ice. Velocity varies across a glacier as well as with depth and these variations cause **crevasses** to form. Glaciers move in three main ways - **basal slide**, **internal deformation**, and **compression and extension** of the ice surface.

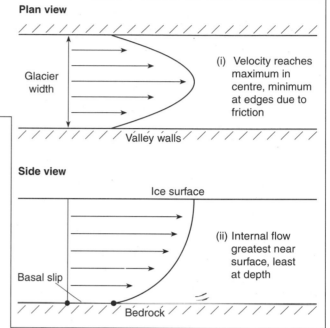

Plan view

Glacier width

(i) Velocity reaches maximum in centre, minimum at edges due to friction

Valley walls

Side view

Ice surface

Basal slip

Bedrock

(ii) Internal flow greatest near surface, least at depth

BASAL SLIDE

Regelation flow

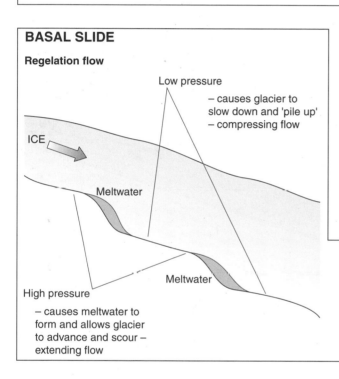

Low pressure

– causes glacier to slow down and 'pile up'
– compressing flow

ICE

Meltwater

Meltwater

High pressure

– causes meltwater to form and allows glacier to advance and scour – extending flow

A thin film of meltwater at the base of warm glaciers allows it to slide over the bedrock. This accounts for up to 75% of glacier movement. Part of the movement is caused by **regelation**, the melting and subsequent re-freezing of water around irregularities in the valley floor. The increase of pressure upstream of the irregularity causes the melting to take place.

INTERNAL DEFORMATION

This involves the movement of ice crystals as a result of gravity. Ice acts like plasticine: on a horizontal surface it remains intact but when suspended at an angle it warps. Internal deformation is especially acute when gradients are high.

Individual crystals may move relative to other crystals (intergranular flow) or along layers - laminar flow - within the glacier.

EXTENDING AND COMPRESSING FLOW

Ice cannot deform rapidly. Consequently it fractures and movement takes place along a series of planes causing crevasses to form.

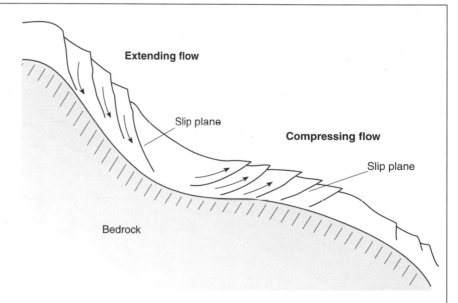

Extending flow

Slip plane

Compressing flow

Slip plane

Bedrock

Glacial erosion

GLACIAL EROSION: PROCESSES AND CONTROLS

The amount and rate of erosion depends on (a) the local geology, (b) the velocity of the glacier, (c) the weight and thickness of the ice, and (d) the amount and character of the load carried. The methods of glacial erosion include plucking and abrasion.

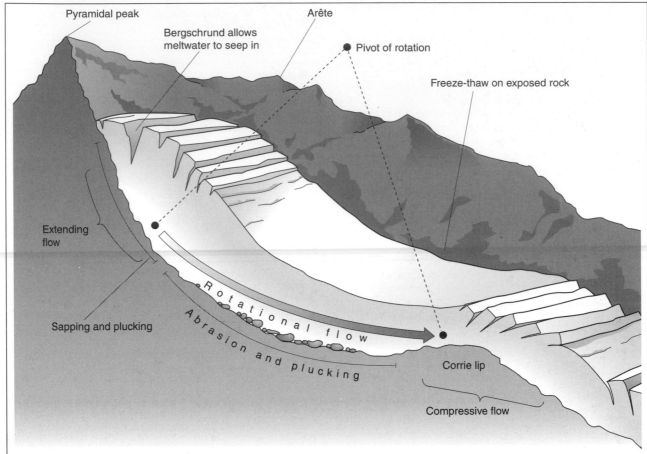

PLUCKING AND ABRASION

Plucking occurs mostly at the base of the glacier and to an extent at the side. It is most effective in jointed rocks or those weakened by sapping (freeze-thaw). As the ice moves, meltwater seeps into the joints and freezes onto the rock, which is then ripped out by the moving glacier.

Abrasion is where the debris carried by the glacier scrapes and scratches the rock leaving **striations**.

Other mechanisms include meltwater, freeze-thaw weathering, and pressure release. Although not strictly glacial nor erosional, these processes are crucial in the development of glacial scenery.

CIRQUES

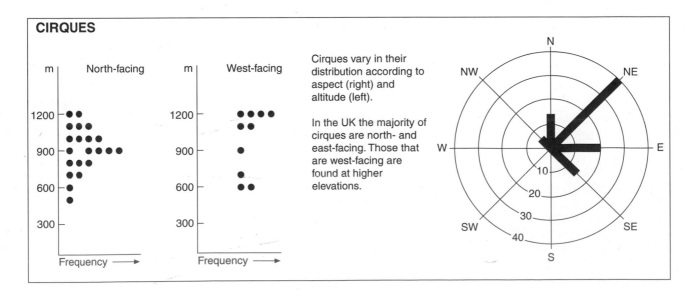

Cirques vary in their distribution according to aspect (right) and altitude (left).

In the UK the majority of cirques are north- and east-facing. Those that are west-facing are found at higher elevations.

Landforms produced by glacial erosion

CIRQUES, ARÊTES, HORNS, AND U-SHAPED VALLEYS

In the UK **cirques** are generally found on north- or east-facing slopes where accumulation is highest and ablation is lowest. They are formed in stages: (i) a pre-glacial hollow is enlarged by **nivation** (freeze-thaw and removal by snow melt); (ii) ice accumulates in the hollow; (iii) having reached a critical weight and depth, the ice moves out in a rotational manner, eroding the floor by plucking and abrasion; (iv) meltwater trickles down the bergschrund allowing the cirque to grow by freeze-thaw. After glaciation, an armchair-shaped hollow remains, frequently filled with a lake, e.g. Red Tarn in the Lake District.

Other features of glacial erosion include **arêtes** and **pyramidal peaks (horns)** caused by the headward recession (cutting back) of two or more cirques. Glacial troughs (or **U-shaped valleys**) have steep sides and flat floors. In plan view they are straight since they have truncated the interlocking spurs of the pre-glacial valley. The ice may also carve deep rock basins frequently filled with **ribbon lakes**. **Hanging valleys** are formed by tributary glaciers which, unlike rivers, do not cut down to the level of the main valley, but are left suspended above, e.g. Stickle Beck. They are usually marked by waterfalls.

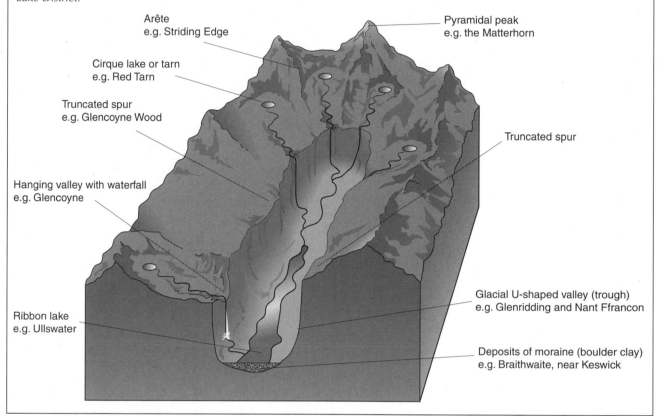

Arête
e.g. Striding Edge

Cirque lake or tarn
e.g. Red Tarn

Truncated spur
e.g. Glencoyne Wood

Hanging valley with waterfall
e.g. Glencoyne

Ribbon lake
e.g. Ullswater

Pyramidal peak
e.g. the Matterhorn

Truncated spur

Glacial U-shaped valley (trough)
e.g. Glenridding and Nant Ffrancon

Deposits of moraine (boulder clay)
e.g. Braithwaite, near Keswick

CRAG AND TAIL

A **crag and tail** is formed when a very large resistant object obstructs ice flow. The ice is forced around the obstruction, eroding weaker rock. Material immediately in the lee of the obstruction is protected by the crag and forms a tail. Edinburgh Castle rock is an ancient volcanic plug whereas its tail is formed of limestone.

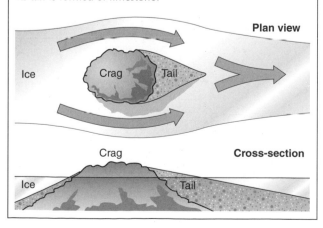

Plan view

Ice   Crag   Tail

Crag   **Cross-section**

Ice   Tail

ROCHES MOUTONNÉES

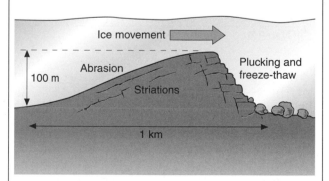

Ice movement

Abrasion   Plucking and freeze-thaw

100 m   Striations

1 km

Roches moutonnées vary in size from a few metres to hundreds of metres. They are smoothed and polished on the up-valley side (stoss) by abrasion, but plucked on the lee side (down valley) as the ice accelerates. They can be over 100 m in height and several kilometres long. In Scotland, the rocky ridges are interspersed with small basins filled with water giving a **cnoc and lochan** landscape (hillock and lake).

Glacial deposition

The term **drift** refers to all glacial and fluvioglacial deposits left after the ice has melted. Glacial deposits or **till** are angular and unsorted, and their long axes are orientated in the direction of glacier flow. These include erratics, drumlins, and moraines. Till is often subdivided into **lodgement till**, material dropped by actively moving glaciers, and **ablation till**, deposits dropped by stagnant or retreating ice.

DEPOSITIONAL FEATURES

Erratics
Erratics are large boulders foreign to the local geology, e.g. the Bowder Stone in Borrowdale and the Norber Stone on the North Yorks Moors.

Moraines
Moraines are lines of loose rocks, weathered from the valley sides and carried by the glaciers. At the snout of the glacier is a crescent-shaped mound of **terminal moraine**. Its character is determined by the amount of load the glacier was carrying, the speed of movement, and the rate of retreat. The ice-contact slope (up-valley) is always steeper than the down-valley slope. The finest example in Britain is the Cromer Ridge, up to 90 m high and 8 km wide.

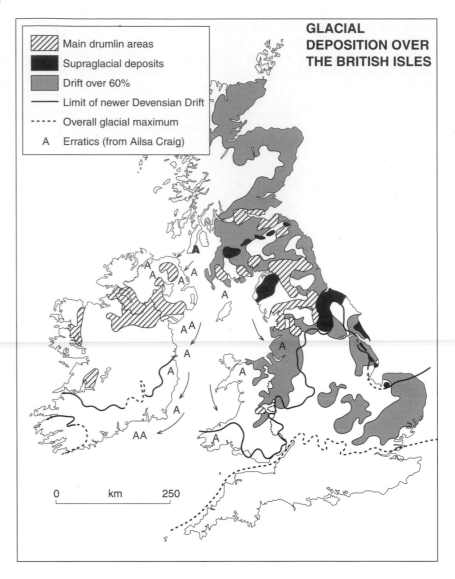

GLACIAL DEPOSITION OVER THE BRITISH ISLES

Legend:
- Main drumlin areas
- Supraglacial deposits
- Drift over 60%
- Limit of newer Devensian Drift
- - - - Overall glacial maximum
- A Erratics (from Ailsa Craig)

0   km   250

Drumlins
Drumlins are small oval mounds up to 1.5 km long and 100 m high, e.g those in the Ribble Valley or the drowned drumlins of Clew Bay in Co. Mayo, Ireland. They are deposited due to friction between the ice and the underlying geology, causing the glacier to drop its load. As the glacier continues to advance it streamlines the mounds.

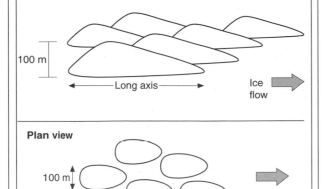

100 m

Long axis

Ice flow

Plan view

100 m

1 km

These features can be used to determine the **direction of glacier movement**. Erratics pinpoint the origin of the material, and drumlins and the long axes of pebbles in glacial till are orientated in the direction of glacier movement.

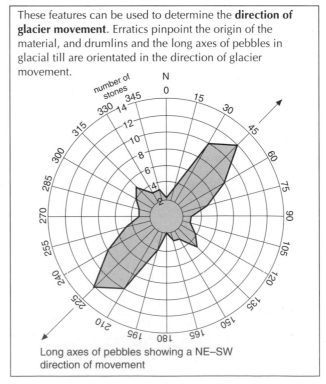

Long axes of pebbles showing a NE–SW direction of movement

Lowland glaciation

Fluvioglacial erosion also has a pronounced impact on the landscape. **Proglacial** lakes are formed adjacent to the ice mass. They may be trapped by high ground or between a retreating ice mass and terminal moraine. **Cols** or spillways are channels formed by meltwater in highland areas when it drains through the lowest part of a watershed. A series of proglacial lakes in the North Yorks Moors were envisaged, where meltwater and rivers were trapped between high ground and the advancing ice mass, thereby creating massive lakes such as Lake Pickering. These grew until they reached a low point in the surrounding landscape, and then drained through in huge meltwater channels, such as Newtondale. However, this theory has been challenged by others claiming the lakes were not as large and that the channels were either marginal meltwater channels or subglacial channels.

Fluvioglacial or **meltwater** deposits can be subdivided into **prolonged drift**, in which the material is very well sorted, e.g. varves and outwash plains, and **ice-contact stratified drift**, e.g. kames and eskers, which are more varied in character.

Kames

Kames are irregular mounds of sorted sands and gravels, formed by supraglacial streams on stagnating ice sheets. Often they contain **kettle holes**, caused by the deposition of material around broken blocks of ice. **Kame terraces** are found at the side of the valley, laid down by streams occupying the site between the valley wall and the glacier, e.g. the Lammermuir Hills in eastern Scotland

Eskers

Eskers are elongated ridges of coarse, stratified, fluvioglacial sands and gravels. Two explanations are given for their formation: (i) material is deposited in subglacial meltwater tunnels or (ii) eskers may represent a rapidly retreating delta, formed as the ice melts and subglacial streams are suddenly released of pressure.

Varves

Varves or **varved clays** are layered deposits of alternating coarse and fine material. Each varve is a year's deposit: coarser material is carried in the lakes by spring meltwaters whereas finer material does not settle until later in the autumn.

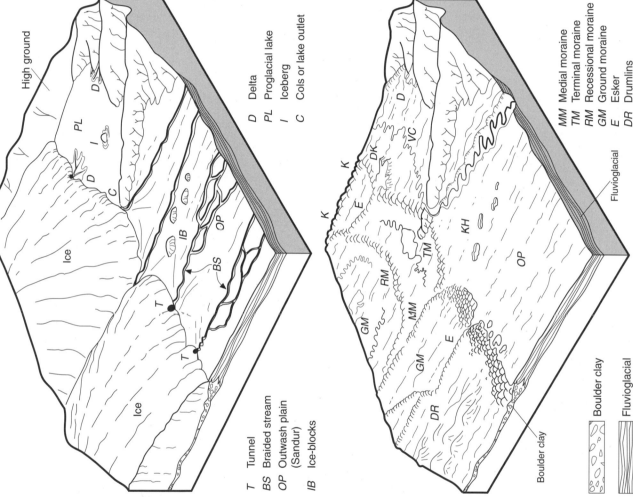

D	Delta
PL	Proglacial lake
I	Iceberg
C	Cols or lake outlet
T	Tunnel
BS	Braided stream
OP	Outwash plain (Sandur)
IB	Ice-blocks

MM	Medial moraine
TM	Terminal moraine
RM	Recessional moraine
GM	Ground moraine
E	Esker
DR	Drumlins
K	Kame
KH	Kettle hole
VC	Varved clay
DK	Delta kame

Boulder clay

Fluvioglacial

FLUVIOGLACIAL LANDFORMS

Outwash plains

Sandur or **outwash plains** are gently sloping plains comprised of sands and gravels that are sorted and stratified. The coarser gravels are deposited first owing to a reduction of meltwater competence, and closer to the ice margin, whereas the sands are carried further down the plain. There is also vertical layering as well as horizontal stratification. These are characterised by braided rivers, heavily laden with sand and gravel, varying enormously in seasonal discharge, causing numerous islets and channels to be formed, e.g. the River Eyra in Iceland.

Periglaciation

PERIGLACIAL ENVIRONMENTS

Periglacial areas are found on the edge of glaciers or ice masses and are characterised by **permafrost** and **freeze-thaw** action. Summer temperatures rise above freezing so ice melts. Three types of periglacial region can be identified: Arctic continental, Alpine, and Arctic Maritime. These vary in terms of mean annual temperature and therefore the frequency and intensity with which processes operate.

Periglacial environments extended during glacial phases. Much of southern Britain, especially those areas not covered by glaciers, were exposed to periglacial processes for considerable lengths of time. Such effects were most marked on rocks such as chalk and limestone which underwent changes in permeability during periglacial phases. Periglacial features may also be found in glaciated areas, formed during periods of glacial wasting (deglaciation). In Scotland, the periglacial zone expanded 11,000 years ago as a result of the Loch Lomond Readvance.

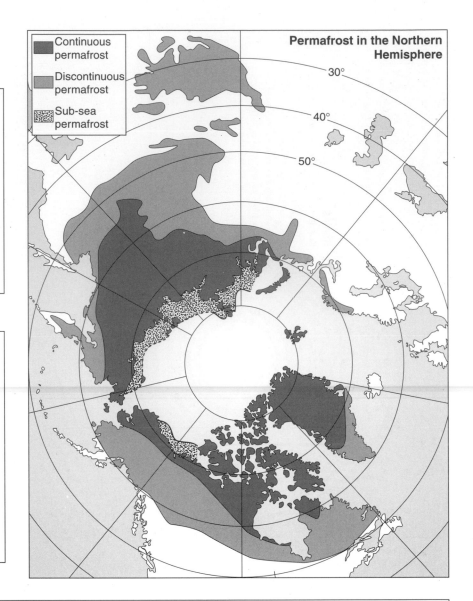

Permafrost in the Northern Hemisphere

Legend:
- Continuous permafrost
- Discontinuous permafrost
- Sub-sea permafrost

30°, 40°, 50°

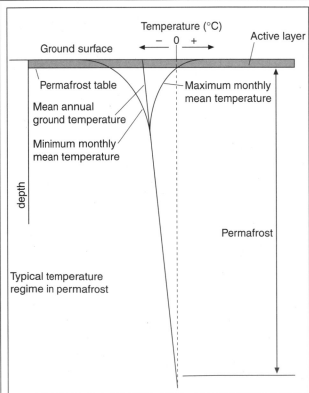

Temperature (°C)
− 0 +

Active layer

Ground surface

Permafrost table

Mean annual ground temperature

Minimum monthly mean temperature

Maximum monthly mean temperature

depth

Permafrost

Typical temperature regime in permafrost

PERMAFROST

Periglacial areas are also associated with **permafrost**, impermeable permanently frozen ground. Approximately 20% of the world's surface is underlain by permafrost, in places up to 700 m deep. Three types of permafrost exist - **continuous**, **discontinuous**, and **sporadic** - and these are associated with mean annual temperatures of –5° to –50°C, –1.5° to –5°C, and 0° to –1.5°C respectively. Above the permafrost is found the **active layer**, a highly mobile layer which seasonally thaws out and is associated with intense mass movements. The depth of the active layer depends upon the amount of heat it receives and varies in Siberia from 0.2-1.6 m at 70°N and from 0.7-4 m at 50°N.

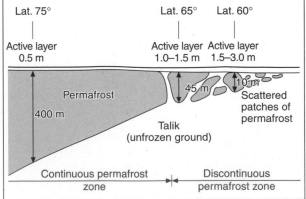

Lat. 75°
Active layer 0.5 m

Lat. 65°
Active layer 1.0–1.5 m

Lat. 60°
Active layer 1.5–3.0 m

Permafrost 400 m

45 m

10 m

Talik (unfrozen ground)

Scattered patches of permafrost

Continuous permafrost zone

Discontinuous permafrost zone

Periglacial processes

Periglacial environments are dominated by **freeze-thaw** weathering. This occurs as temperatures fluctuate above and below freezing point. As water freezes it expands by 10%, exerting pressures of up to 2100 kg/cm². Most rocks can only withstand up to 210 kg/cm². It has most effect on well jointed rocks which allow water to seep into cracks and fissures. **Congelifraction** refers to the splitting of rocks by freeze-thaw action.

Solifluction (**gelifluction** or **congelifluction**) literally means flowing soil. In winter, water freezes in the soil causing expansion of the soil and segregation of individual soil particles. In spring, the ice melts and water flows downhill. It cannot infiltrate into the soil because of the impermeable permafrost. As it moves over the permafrost it carries segregated soil particles (peds) and deposits them further downslope as a solifluction lobe or terracette.

Cambering is the process whereby segments of rock become dislodged from the main body of rock and begin to move downhill. It is aided by freeze-thaw.

Nivation (**altiplanation** or **cryoplanation**) is freeze-thaw weathering under a snow bank. The broken material is removed in spring and summer by the melted snow.

Chemical weathering is also effective in periglacial regions. **Carbonation** is an important process because of the low temperatures. Carbon dioxide is more soluble at low temperature hence the water becomes quite acidic. It is aided by the slowly rotting vegetation which releases organic acids. **Hydrolysis** is also important because of the presence of organic acids in the marshy soil.

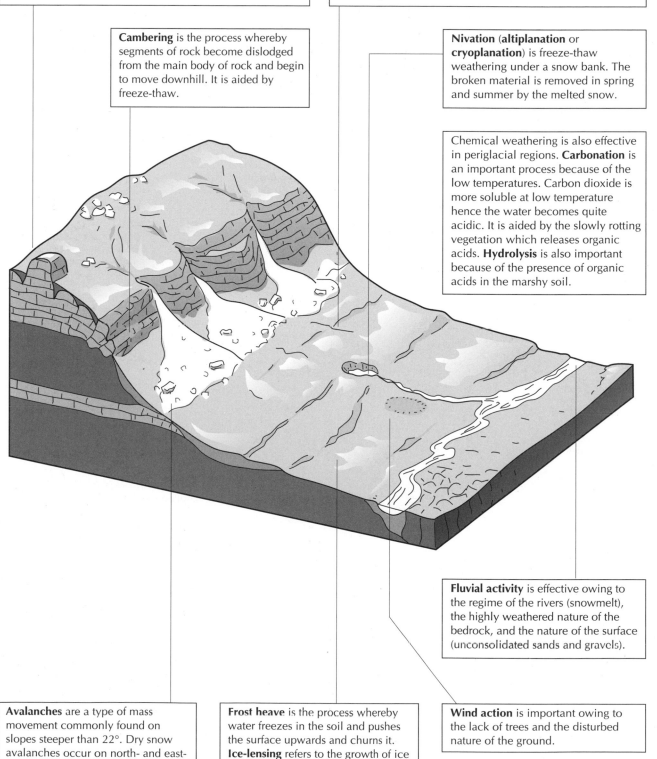

Fluvial activity is effective owing to the regime of the rivers (snowmelt), the highly weathered nature of the bedrock, and the nature of the surface (unconsolidated sands and gravels).

Avalanches are a type of mass movement commonly found on slopes steeper than 22°. Dry snow avalanches occur on north- and east-facing slopes, where the snow is unstable, whereas wet snow flows generally result from rapid snow melt.

Frost heave is the process whereby water freezes in the soil and pushes the surface upwards and churns it. **Ice-lensing** refers to the growth of ice crystals in soil. **Geliturbation** and **congeliturbation** are other terms for frost heave.

Wind action is important owing to the lack of trees and the disturbed nature of the ground.

Periglacial landforms

Tors are isolated outcrops of bare rock, e.g. Yes Tor and Hay Tor on Dartmoor. They are formed as a result of intense frost shattering under periglacial conditions and the removal of the weathered material (growan) by mass movements and fluvial activity (the Palmer-Nielson theory).

Scree slopes are slopes composed of large quantities of angular fragments of rock, e.g. the slopes at Wastwater in the Lake District. Typically they have an angle of rest of about 35°. Extensive upland surfaces of angular rocks are known as **blockfields.**

Dry valleys are river valleys without rivers. They are most commonly found on chalk and limestone such as The Manger at Uffington (Vale of the White Horse) and the Devil's Dyke near Brighton. During the periglacial period, limestone and chalk became impermeable owing to permafrost, and therefore rivers flowed over their surfaces. High rates of fluvial erosion occurred because of spring melt, the highly weathered nature of the surface, and high rates of carbonation. At the end of the periglacial period normal permeability returned, waters sank into the permeable rocks, and the valleys were left dry.

Patterned ground is a general term describing the stone circles, polygons, and stripes that are found in soils subjected to intense frost action, e.g. Grimes Graves near Thetford in Norfolk. On steeper slopes, stone stripes replace stone circles and polygons. Their exact mode of formation is unclear although ice sorting, differential frost heave, solifluction, and the effect of vegetation are widely held to be responsible.

Rivers in periglacial areas are typically **braided**, with numerous small channels separated by small linear islands, e.g. the River Eyra in Iceland. Braiding occurs because the river is carrying too much sediment as a result of the highly erosive nature of streams fed by spring melt, and is forced to deposit some.

Loess refers to deposits laid down by the wind. They consist mostly of unstratified, structureless silt, and cover extensive areas in China and northern Europe and produce smaller deposits in Britain such as the Brickearth of East Anglia, e.g. Wangford Warren.

Solifluction **terracettes**, like those at Maiden Castle in Dorset, are step-like features ('sheep-walks') caused by mass movement in the active layer. **Solifluction lobes** are elongated versions typically 20 m long and over 1 m high.

Asymmetric slopes are valleys with differing slope angles, e.g. the River Exe in Devon. They are caused by variations in **aspect** which affects the frequency and intensity of weathering. South-facing slopes receive more insolation and are subjected to more mass wasting, and hence have lower slope angles. By contrast, north-facing slopes remain frozen for longer periods, are not as highly weathered, and consequently remain steeper.

Coombe rock or **head** is a periglacial deposit comprising of chalk, mud, and/or clay, compacted with angular fragments of frost shattered rock, e.g. at Scratchey Bottom near Durdle Door in Dorset.

A **pingo** is an isolated, conical hill up to 90 m high and 800 m wide, which can only develop in periglacial areas. They form as a result of the movement and freezing of water under pressure. Two types are generally identified - **open system** and **closed system** pingos. Where the water is from a distant elevated source, open system pingos are formed, whereas if the supply of water is local, and the pingo occurs as a result of the expansion of permafrost, closed system pingos are formed. Nearly 1500 pingos are found in the Mackenzie Delta of Canada, and examples of relict pingos can be found in the Vale of Llanberris in Wales. When a pingo collapses ramparts and ponds are left.

Problems in the use of periglacial areas

The hazards associated with the use of periglacial areas are diverse and may be intensified by human impact. Problems include mass movements such as avalanches, solifluction, rockfalls, frost heave, and icings as well as flooding, thermokarst subsidence, low temperatures, poor soils, a short growing season, and a lack of light.

For example, the Nyenski tribe in the Yamal Peninsula of Siberia have suffered as a result of the exploitation of oil and gas. Oil leaks, subsidence of railway lines, destruction of vegetation, decreased fish stocks, pollution of breeding grounds, and reduced caribou numbers have all happened directly or indirectly as a result of man's attempt to exploit this remote and inhospitable environment.

Close to rivers, **frost heave** is very significant (owing to an abundant supply of water) and can lift piles and structures out of the ground. **Piles** for carrying oil pipelines therefore need to be embedded deep in the permafrost to overcome mass movement in the active layer. In Prudhoe Bay, Alaska, they are 11 m deep. However, this is extremely expensive: each one cost over $3000 in the early 1970s.

Services are difficult to provide in periglacial environments. It is impossible to lay underground networks and so **utilidors**, insulated water and sewage pipes, are provided above ground. Waste disposal is also difficult owing to the low temperatures.

Alpine periglacial areas also suffer environmental pressures. Here the concerns are more than damage to the physical environment, as traditional economies have declined at the expense of electro-chemical and services industries, especially tourism. An elaborate infrastructure is required to cope with the demands of an affluent tourist population, and this may undermine the natural environment and traditional societies.

Traditionally, periglacial pastures have been used by Inuits for herding or hunting caribou. The abundance of lakes allows travel by float plane and the frozen winter rivers and lack of trees enables overland travel. Periglacial areas are **fragile** for two reasons. First, the ecosystem is highly susceptible to interference because of the limited number and diversity of species involved. The extremely low temperatures limit decomposition, and hence **pollution**, especially oil spills, have a very long-lasting effect on periglacial ecosystems. Secondly, **permafrost** is easily disrupted. The disruption of permafrost poses significant problems. Heat from buildings and pipelines, and changes in the vegetation cover, rapidly destroy it. Thawing of the permafrost increases the active layer and the subsequent settling of the soil causes subsidence. Consequently, engineers have either built structures on a bed of gravel, up to 1 m thick for roads, or used stilts.

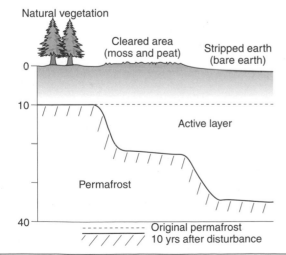

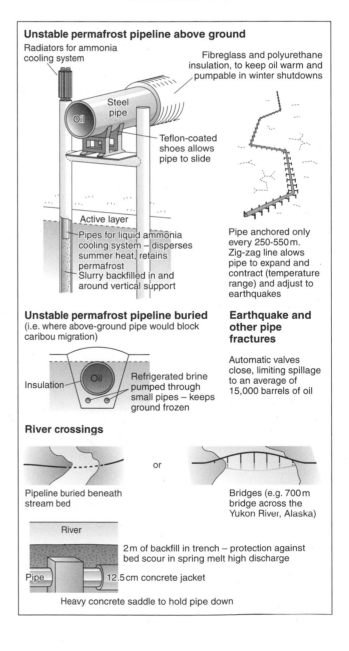

Waves and tides

WAVE TERMINOLOGY

Wavelength or **amplitude** is the distance between two successive crests or troughs.

Wave period is the time in seconds between two successive crests or troughs.

Wave frequency is the number of waves per minute.

Wave height is the distance between the trough and the crest.

The **fetch** is the amount of open water over which a wave has passed.

Velocity is the speed a wave travels at, and is influenced by wind, fetch, and depth of water.

Swash is the movement of water up the beach.

Backwash is the movement of water down the beach.

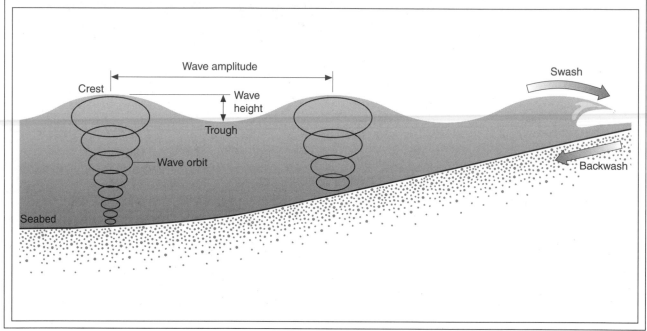

DESTRUCTIVE WAVES

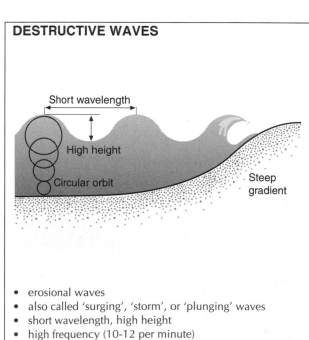

- erosional waves
- also called 'surging', 'storm', or 'plunging' waves
- short wavelength, high height
- high frequency (10-12 per minute)
- circular orbit
- low period (one every 5-6 seconds)
- backwash greater than swash
- steep gradient

CONSTRUCTIVE WAVES

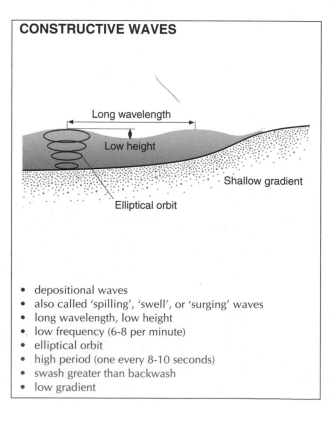

- depositional waves
- also called 'spilling', 'swell', or 'surging' waves
- long wavelength, low height
- low frequency (6-8 per minute)
- elliptical orbit
- high period (one every 8-10 seconds)
- swash greater than backwash
- low gradient

WAVE REFRACTION AND LONGSHORE DRIFT

Wave *refraction* occurs when waves approach an irregular coastline or at an oblique angle (a). Refraction reduces wave velocity and, if complete, causes wave fronts to break parallel to the shore. Wave refraction concentrates energy on the flanks of headlands and dissipates energy in bays (b). However, refraction is rarely complete, and consequently *longshore drift* occurs (c).

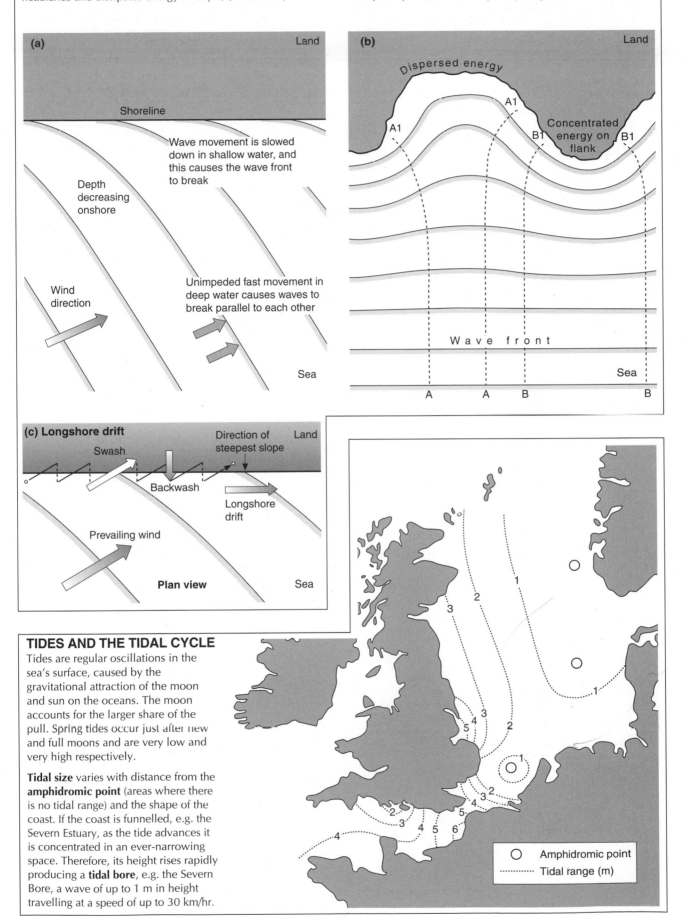

(a)
Land
Shoreline
Wave movement is slowed down in shallow water, and this causes the wave front to break
Depth decreasing onshore
Wind direction
Unimpeded fast movement in deep water causes waves to break parallel to each other
Sea

(b)
Land
Dispersed energy
A1
A1
Concentrated energy on flank
B1
B1
Wave front
Sea
A   A   B   B

(c) Longshore drift
Land
Swash
Direction of steepest slope
Backwash
Longshore drift
Prevailing wind
Plan view
Sea

TIDES AND THE TIDAL CYCLE

Tides are regular oscillations in the sea's surface, caused by the gravitational attraction of the moon and sun on the oceans. The moon accounts for the larger share of the pull. Spring tides occur just after new and full moons and are very low and very high respectively.

Tidal size varies with distance from the **amphidromic point** (areas where there is no tidal range) and the shape of the coast. If the coast is funnelled, e.g. the Severn Estuary, as the tide advances it is concentrated in an ever-narrowing space. Therefore, its height rises rapidly producing a **tidal bore**, e.g. the Severn Bore, a wave of up to 1 m in height travelling at a speed of up to 30 km/hr.

○ Amphidromic point
...... Tidal range (m)

Coastal erosion

PROCESSES

- **Abrasion** The wearing away of the shoreline by material carried by the waves.
- **Hydraulic impact** The force of water and air on rocks (up to 30,000 kg/m² in severe storms).
- **Solution** The wearing away of base-rich rocks, especially limestone, by acidic water. Organic acids aid the process.
- **Attrition** The rounding and reduction of particles carried by the waves.

Additionally, sub-aerial processes such as mass movements, wind erosion, and weathering are important.

CLIFF SHAPES

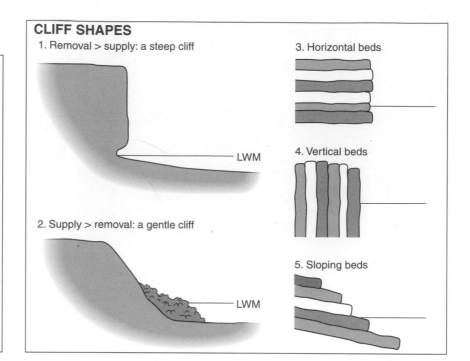

1. Removal > supply: a steep cliff — LWM

2. Supply > removal: a gentle cliff — LWM

3. Horizontal beds

4. Vertical beds

5. Sloping beds

EVOLUTION OF SHORE PLATFORMS

A model of cliff and shore platform evolution shows how a steep cliff (1) is replaced by a lengthening platform and lower angle cliff (5) subjected to sub-aerial processes rather than marine forces.

Alternatively, platforms might be formed by (i) frost action, (ii) salt weathering, or (iii) biological action during lower sea-levels and different climates.

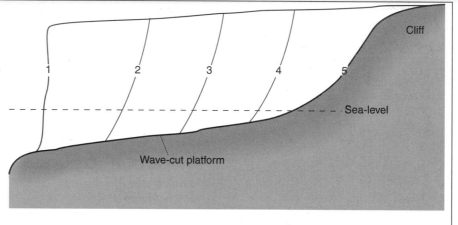

Cliff

1 2 3 4 5

Sea-level

Wave-cut platform

FEATURES OF COASTAL EROSION

Stump	Stack	Arch	Headland
Bull, Cow, and Calf rocks	Man o'War rocks	Durdle Door	Durdle Promontory

High tide

Low tide

RATES OF CLIFF EROSION

Location	Geology	Erosion (metres per 100 years)
Holderness	Glacial drift	120
Cromer, Norfolk	Glacial drift	96
Folkestone	Clay	28
Isle of Thanet	Chalk	7-22
Seaford Head	Chalk	126
Beachy Head	Chalk	106
Barton, Hants	Barton Beds	58

Erosion is highest when there are:
- frequent storm waves
- easily erodable material

Coastal deposition

BEACH PROFILE

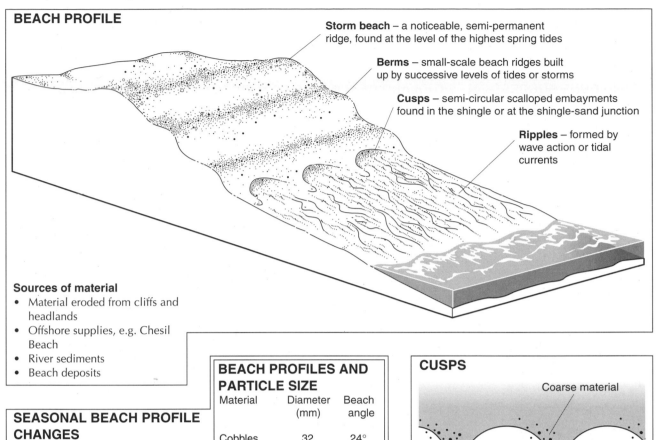

Storm beach – a noticeable, semi-permanent ridge, found at the level of the highest spring tides

Berms – small-scale beach ridges built up by successive levels of tides or storms

Cusps – semi-circular scalloped embayments found in the shingle or at the shingle-sand junction

Ripples – formed by wave action or tidal currents

Sources of material
- Material eroded from cliffs and headlands
- Offshore supplies, e.g. Chesil Beach
- River sediments
- Beach deposits

SEASONAL BEACH PROFILE CHANGES

Upper sweep profile (summer)

Sweep zone

Lower sweep profile (winter)

High tide (spring)
High neap
Mean sea-level
Low neap
Low tide (spring)

BEACH PROFILES AND PARTICLE SIZE

Material	Diameter (mm)	Beach angle
Cobbles	32	24°
Pebbles	4	17°
Coarse sand	2	7°
Medium sand	0.2	5°
Fine sand	0.02	3°
Very fine sand	0.002	1°

CUSPS

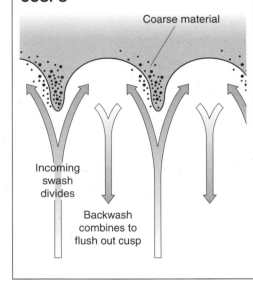

Coarse material

Incoming swash divides

Backwash combines to flush out cusp

FEATURES OF DEPOSITION

Essential requirements include:
- a large supply of material
- longshore drift
- an irregular, indented coastline, e.g. river mouths

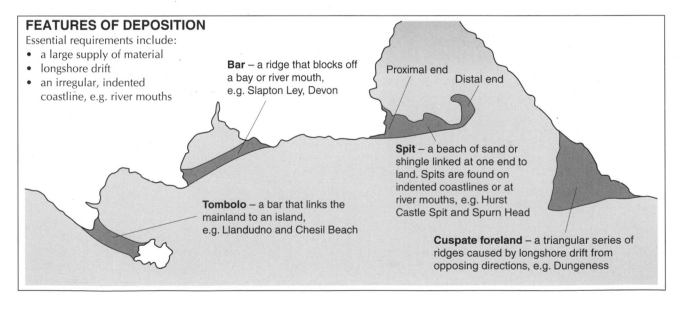

Bar – a ridge that blocks off a bay or river mouth, e.g. Slapton Ley, Devon

Proximal end

Distal end

Spit – a beach of sand or shingle linked at one end to land. Spits are found on indented coastlines or at river mouths, e.g. Hurst Castle Spit and Spurn Head

Tombolo – a bar that links the mainland to an island, e.g. Llandudno and Chesil Beach

Cuspate foreland – a triangular series of ridges caused by longshore drift from opposing directions, e.g. Dungeness

Coastal ecosystems

An ecosystem is a set of inter-related plants and animals with their non-living environment. Coastal ecosystems include sand dunes, **psammoseres**, and salt marshes, **haloseres**.

These change spatially and temporally. The changes in micro-environment which allow other species to invade, compete, succeed, and dominate is termed **succession**.

SAND DUNE SUCCESSION - STUDLAND BEACH, ISLE OF PURBECK

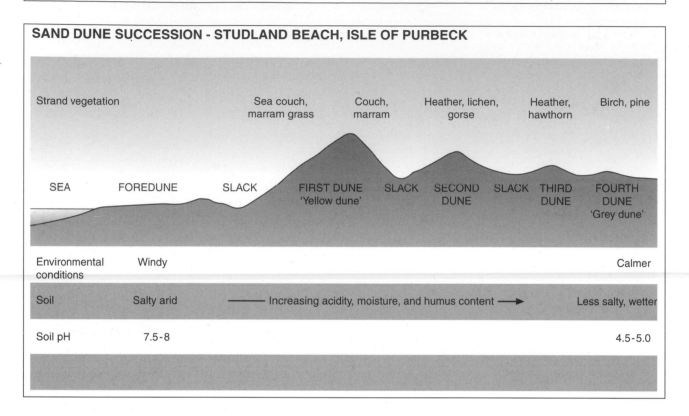

	Strand vegetation		Sea couch, marram grass	Couch, marram	Heather, lichen, gorse	Heather, hawthorn	Birch, pine
	SEA	FOREDUNE	SLACK	FIRST DUNE 'Yellow dune'	SLACK — SECOND DUNE — SLACK	THIRD DUNE	FOURTH DUNE 'Grey dune'
Environmental conditions	Windy						Calmer
Soil	Salty arid		⟶ Increasing acidity, moisture, and humus content ⟶				Less salty, wetter
Soil pH	7.5-8						4.5-5.0

SALT MARSHES

Examples include Scolt Head Island in East Anglia and Newtown on the Isle of Wight.

Salt marshes are very productive and fertile ecosystems because of their high oxygen content, nutrient availability, and light availability, and because of the cleaning action of the tides.

I **Colonisers** on bare mud flats: algae (enteromorpha), eel grass, and marsh samphire (salicornia) increase the amount of deposition of silt. These plants can tolerate alkaline conditions and regular inundation by sea-water.

II **Halophytic vegetation** such as rice grass and spartina (cord grass) build up the salt marsh by as much as 5 cm per annum. Their roots anchor into the soft mud; the vegetation is taller and longer living than salicornia but not as salt tolerant.

III Sea lavendar **grasses**: inundated only at spring tides. Less salt tolerant.

IV A **raised salt marsh** with creeks may be formed, including turf grasses such as fescue and rushes (juncus). Inundation is rare.

V Inundation absent: ash and alder.

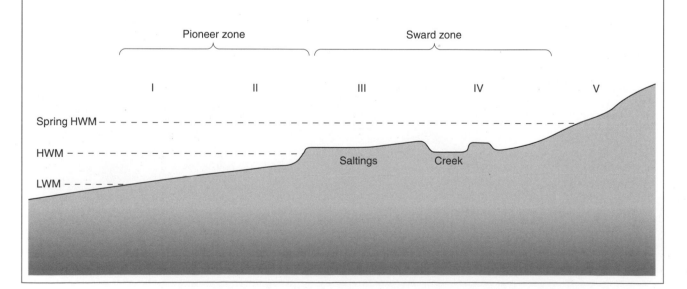

Coastal classification

SEA-LEVEL CHANGES

Sea-levels change in connection with the growth and decay of ice sheets. **Eustatic** change refers to a global change in sea-level. The level of the land also varies in relation to the sea. Land may rise as a result of tectonic uplift or following the removal of an ice sheet. The change in the level of the land relative to the level of the sea is known as **isostatic adjustment** or **isostacy**.

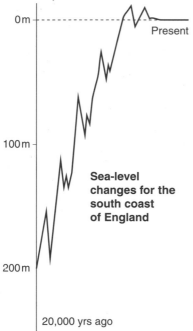

Sea-level changes for the south coast of England

Features of emerged coastlines include:
- raised beaches, e.g. Portland raised beach
- coastal plains
- relict cliffs, e.g. the Fall Line in eastern USA

Submerged coastlines include:
- rias, e.g. the River Fal, a drowned river valley
- fjords, e.g. Loch Torridan, a drowned U-shaped valley
- fjards or drowned glacial lowlands

Coasts can be classified in a number of ways
- high or low **energy** coastlines
- erosional or depositional
- submerged or emerged
- macro- meso- or micro-tidal
- storm- or swell - wave environment
- advancing or retreating coastlines

These are mainly descriptive categories apart from Valentin's classification which links the first two factors. However, the categories overlap - high energy coasts, for example, are mainly erosional or storm wave coasts.

TIDAL CLASSIFICATION
- Macro-tidal   > 4 m
- Meso-tidal   2 - 4 m
- Micro-tidal   2 m

WAVE ENVIRONMENTS

Storm wave environments, e.g. the British Isles and mid-latitude coast-lines dominated by waves generated thousands of kilometres away.

Swell wave environments, e.g. the tropical trade wind areas dominated by more gentle winds.

VALENTIN'S CLASSIFICATION

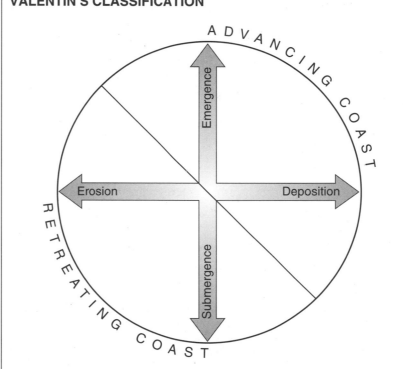

- **Retreating coasts** include submerged coats and coasts where the rate of erosion is greater than the rate of emergence.
- **Advancing coasts** include emerged coastlines and coasts where deposition is rapid.

ATLANTIC AND PACIFIC COASTLINES

This classification is based on whether the trend (direction) of the geology is parallel to or at right angles to the coastline. In **Atlantic** coastlines, the geology is at right angles to the coast, e.g. South-West Ireland, whereas in **Pacific** coastlines the geological trend is parallel to the coastline, e.g. California. Both types of coastline are illustrated by the example of the Isle of Purbeck.

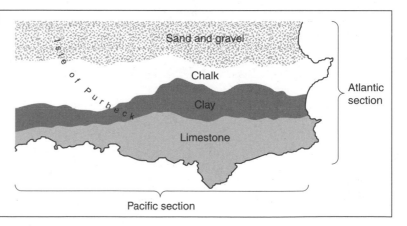

Coastal management

There are a number of ways in which people affect coastal systems, e.g. by dredging and extracting sediment, and building flood protection schemes. Attempts to protect the coast in one place may increase pressure elsewhere, as is often the case with groynes. The best protection for a coast is a beach. However, many beaches are **relict** features, i.e. they have stopped forming as their sediment supply has run out.

LONGSHORE DRIFT

Groynes can prevent longshore drift from removing a beach by interrupting the natural flow of sediment. However, by trapping sediment they deprive another area, down-drift, of beach replenishment. Without its beach a coast is increasingly vulnerable to erosion, e.g. the cliffs at Barton on Sea were easily eroded following the construction of groynes along the coast at Bournemouth.

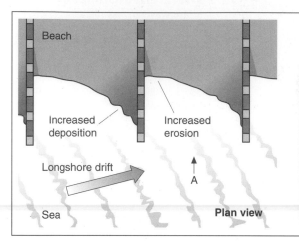

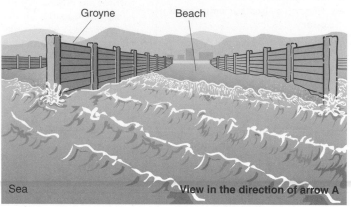

COASTAL FLOODING

This is a serious problem in the low-lying areas of eastern Egland. A combination of factors created severe floods in February 1953, when over 300 people were killed, over 800 km² was flooded, and the cost of the damage exceeded £50 million. These included:

- an intense low pressure system, 970 mb, with winds of over 100 knots
- high tides
- high river levels

SAND DUNE EROSION

Sand dunes are extremely susceptible to erosion both by wind and sea. Trampling accelerates the process. By reducing the vegetation cover the sand or soil is not held and is open to wind erosion, which may lead to the formation of blow-out dunes.

There are a number of ways of stabilising dunes:

- planting marram grass
- building walkways or 'duckboards' to reduce trampling
- planting fences and brushwood to trap sand
- land use zoning to prevent areas from suffering pressure

DREDGING

The fishing village of Hallsands was destroyed in 1917 by a severe storm. The shingle beach, which previously had protected the village, had been removed by contractors building the Naval Dockyard extension at Devonport. Up to 660,000 tons of shingle were removed, lowering the beach by up to 5 m. The material was never replaced, as the shingle was a relict feature deposited about 6000 years ago by rising sea-levels. Up to 6 m of cliff erosion occurred between 1907 and 1957.

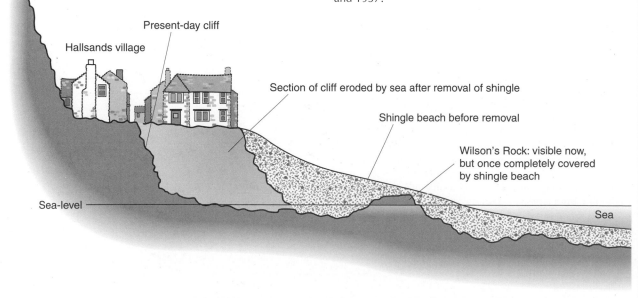

Solutions

SEA-WALLS

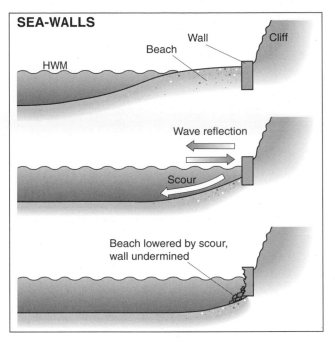

'HARD' ENGINEERING

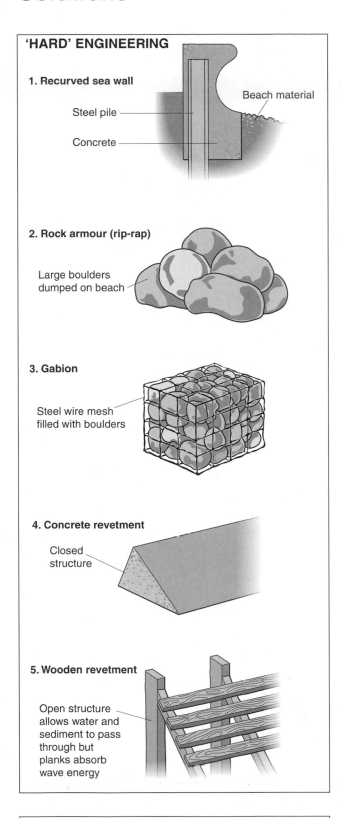

1. Recurved sea wall

Steel pile

Concrete

Beach material

2. Rock armour (rip-rap)

Large boulders dumped on beach

3. Gabion

Steel wire mesh filled with boulders

4. Concrete revetment

Closed structure

5. Wooden revetment

Open structure allows water and sediment to pass through but planks absorb wave energy

MANAGED RETREAT

The cost of protecting Britain's coastline was up to £60 million annually until the early 1990s. Since then government cuts have reduced this. Part of the problem is that southern and eastern England are slowly sinking while sea-level is rising. The risk of flooding and hence the cost of protection are increasing. 'Managed retreat' allows nature to take its course: erosion in some areas, deposition in others. Benefits include less money spent and the creation of natural environments.

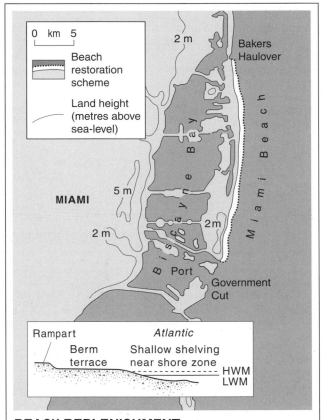

BEACH REPLENISHMENT

Miami is an excellent example of beach replenishment. One of the USA's most popular tourist resorts, by the 1950s there was very little beach left. Erosion threatened the physical, economic, and social future of Miami. Between 1976 and 1982, an 18 km long, 200 m wide beach was constructed using 18 million cubic metres of sand dredged from a zone 3-4 km offshore. As far as possible it replicated a natural beach, although for the benefit of the tourists it has been kept largely free of vegetation. Access for shipping through the barrier island is still necessary, so erosion and drifting of sand still occurs. Approximately 250,000 m³ of sand has to be replenished each year. Although expensive, the high value of tourism, industry, and residential property make this a feasible solution.

The atmosphere

VERTICAL STRATIFICATION

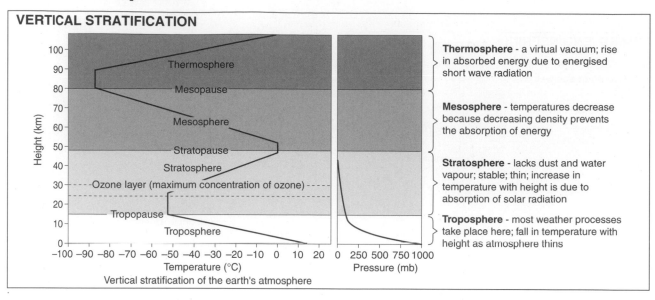

Vertical stratification of the earth's atmosphere

Thermosphere - a virtual vacuum; rise in absorbed energy due to energised short wave radiation

Mesosphere - temperatures decrease because decreasing density prevents the absorption of energy

Stratosphere - lacks dust and water vapour; stable; thin; increase in temperature with height is due to absorption of solar radiation

Troposphere - most weather processes take place here; fall in temperature with height as atmosphere thins

THE EXCHANGE OF ENERGY BETWEEN EARTH AND ATMOSPHERE

Long wave re-radiation
Short wave energy is converted to long wave energy which can be absorbed by the atmosphere; some is re-radiated back. Water vapour acts as an insulator.

Latent heat transfers
Energy is used to convert water into water vapour (evaporation). The heat is retained as latent heat. When vapour turns back into water the heat is released.

Compression heating
When air contracts its temperature increases.

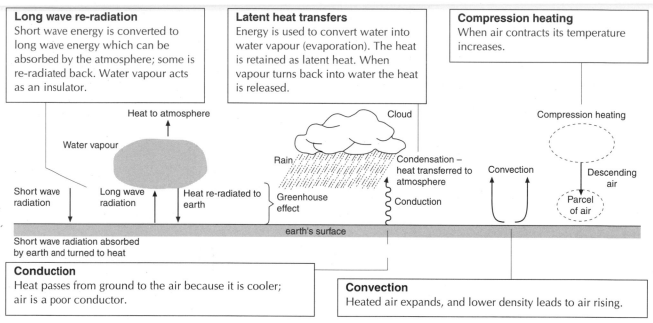

Conduction
Heat passes from ground to the air because it is cooler; air is a poor conductor.

Convection
Heated air expands, and lower density leads to air rising.

TEMPERATURE AND SOLAR RADIATION
The amount of solar radiation received and lost depends on a number of factors:

• **Distance from the sun according to the earth's elliptical orbit**

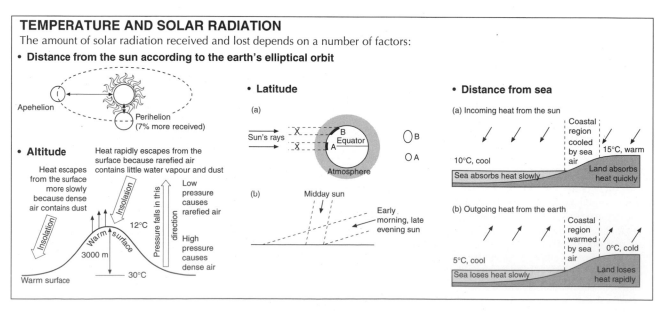

• **Altitude**

• **Latitude**

• **Distance from sea**

Atmospheric motion

Atmospheric motion is controlled by the combination of the following forces:

- pressure-gradient force
- Coriolis force
- the geostrophic wind
- the gradient wind
- friction

WHAT IS THE PRESSURE-GRADIENT FORCE?

The movement of air occurs along pressure gradients from high to low pressure.

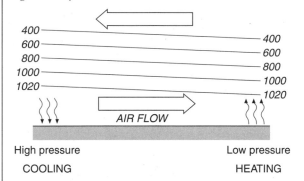

High pressure   Low pressure

COOLING   HEATING

WHAT IS THE CORIOLIS FORCE?

The Coriolis force is the deflection of winds due to the earth's rotation.

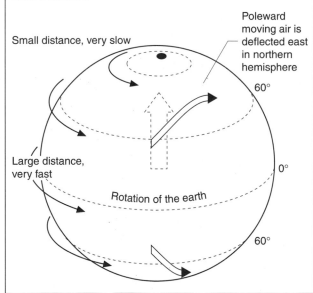

WHAT IS A GRADIENT WIND?

Where isobars are curved, centrifugal and centripetal forces act upon the wind to maintain a flow parallel to the isobars. The curved path is called a gradient wind.

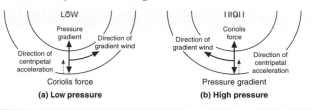

(a) Low pressure   (b) High pressure

HOW WINDS ARE STEERED IN THE UPPER ATMOSPHERE

P   Pressure-gradient force
Co   Coriolis force
C   Centrifugal force
   Isobars
   Wind

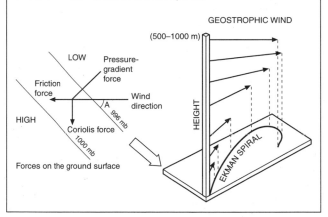

THE EFFECT OF FRICTION

Frictional drag from the earth's surface modifies the balance between horizontal gradient force and the Coriolis force. Friction decreases wind speed but also changes wind direction. This results in Ekman spirals.

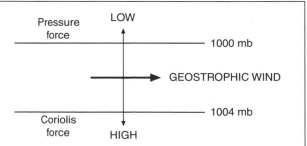

WHAT IS THE GEOSTROPHIC WIND?

In mid-latitudes, the pressure-gradient force and the Coriolis force are directly balanced. This leads to air moving not from high to low pressure but between the two, parallel to the isobars. This is called a geostrophic wind.

The geostrophic wind case of balanced motion (northern hemisphere)

Global circulation models

Low latitudes are warmer than higher latitudes. This energy deficit should result in a large convection cell:

- air rises over the equator due to strong heating

- air then moves polewards to sink

- this is then drawn back to the low pressure

This simple model can be modified in a number of ways.

The Hadley Cell

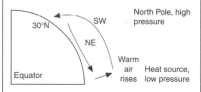

Poleward-flowing currents deflected to the right in the northern hemisphere to become south-westerlies.

The three-cell model

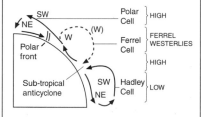

The Hadley Cell model was modified to include three cells in each hemisphere: the Hadley Cell, the Ferrel Cell, and the Polar Cell.

A GENERALISED MODEL OF GLOBAL CIRCULATION

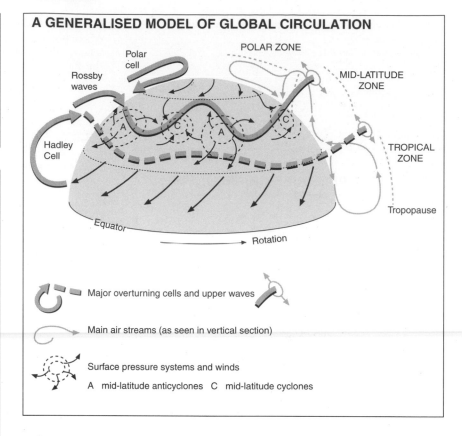

Major overturning cells and upper waves

Main air streams (as seen in vertical section)

Surface pressure systems and winds

A mid-latitude anticyclones   C mid-latitude cyclones

NEW CIRCULATION MODELS

New models change the relative importance of the three convection cells in each hemisphere. These changes are influenced by:

- jet streams - strong and regular winds which blow in the upper atmosphere about 10 km above the surface; they blow between the poles and tropics (100-300 km/h)

- Rossby waves - 'meandering rivers of air' formed by westerly winds; three to six waves in each hemisphere; formed by major relief barriers, thermal differences; uneven land-sea-land interface

The diagram below shows a flow model relating summer convection, the easterly jet stream, and high pressure subsidence over northern Africa and the eastern North Atlantic. It shows the way in which convection cells can be modified by upper air movements.

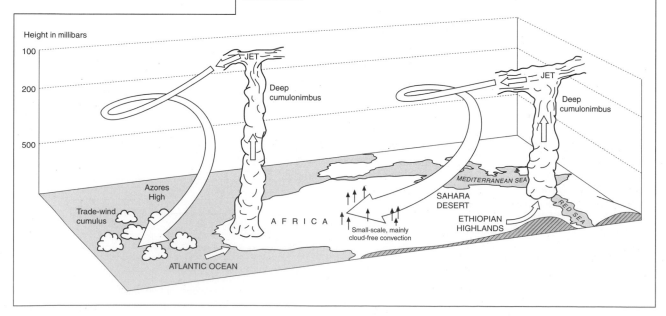

Cloud types

Clouds are classified according to their appearance, form, and height.

THE FOUR MAIN GROUPS OF CLOUDS

A High clouds (6000 – 12000 m)
- Cirrus - composed of small ice crystals (wispy or feather-like).
- Cirrocumulus - ice crystals (globular or rippled).
- Cirrostratus - ice crystals (thin, white, almost transparent sheet which causes the sun and moon to have 'haloes').

B Middle clouds (2100 – 6000 m)
- Altocumulus - water droplets in layers or patches (globular or bumpy-looking).
- Altostratus - water droplets in sheets (grey or watery-looking clouds).

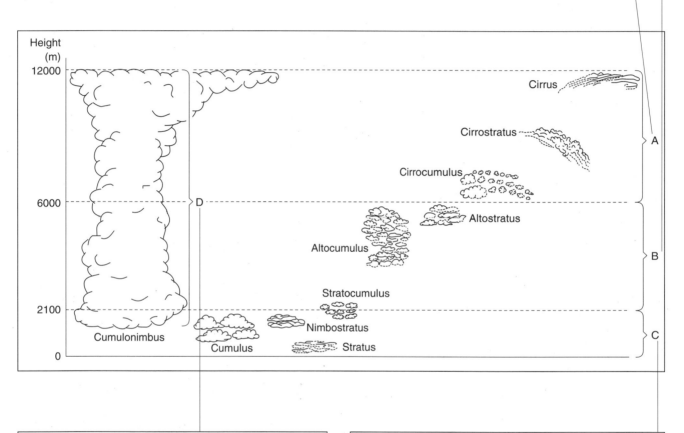

D Clouds with vertical development (1500 – 9000 m)
- Cumulus - round-topped and flat-based (whitish-grey globular mass, individual cloud units).
- Cumulonimbus - great vertical extent (white or black globular masses, rounded tops often spread out in the form of an anvil).

C Low clouds (below 2100 m)
- Stratocumulus - large globular masses (soft and grey with pronounced regular pattern).
- Nimbostratus - dark grey and rainy-looking (dense and shapeless, often leading to regular rain).
- Stratus - low, grey, and layered (bring dull weather and lead to drizzle).

CONDENSATION

Clouds can also be classified according to the mechanism of vertical motion which produces condensation.
Four categories exist:

- gradual uplift of air associated with a low-pressure system
- thermal convection
- uplift by mechanical turbulence (forced convection)
- ascent over an orographic barrier

Precipitation

FORMATION OF PRECIPITATION

There are four conditions needed for the formation of major precipitation (rain, snow, hail):

- air cooling
- condensation and cloud formation
- an accumulation of moisture
- growth of cloud droplets

Two main groups of theories attempt to explain the rapid growth of raindrops:

(i) growth of ice crystals at the expense of water droplets

(ii) coalescence of small water droplets by the sweeping action of falling drops

The growth of ice crystals

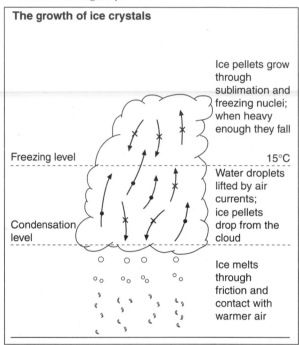

Ice pellets grow through sublimation and freezing nuclei; when heavy enough they fall

Freezing level   15°C

Water droplets lifted by air currents; ice pellets drop from the cloud

Condensation level

Ice melts through friction and contact with warmer air

Collision theories

In the humid tropics, where large amounts of warm, moist air rise rapidly, the raindrops falling from the top of the clouds often collide with water droplets at lower levels, so increasing the size of raindrops.

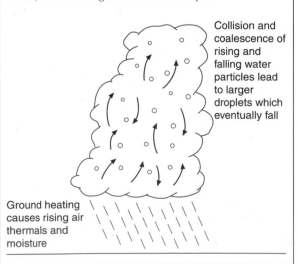

Collision and coalescence of rising and falling water particles lead to larger droplets which eventually fall

Ground heating causes rising air thermals and moisture

TYPES OF RAIN

Convection rain

Humid tropics and the interior of continents in summer; due to strong, upward moving buoyant air; **thunderstorms.**

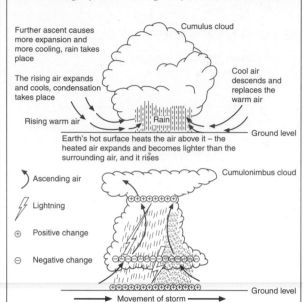

Further ascent causes more expansion and more cooling, rain takes place

Cumulus cloud

The rising air expands and cools, condensation takes place

Cool air descends and replaces the warm air

Rising warm air   Rain   Ground level

Earth's hot surface heats the air above it – the heated air expands and becomes lighter than the surrounding air, and it rises

↗ Ascending air

⚡ Lightning

⊕ Positive change

⊖ Negative change

Cumulonimbus cloud

Ground level

← Movement of storm →

Depression or Cyclonic or Frontal rain

Mid-latitudes; depressions rising over colder air; **consistant rain.**

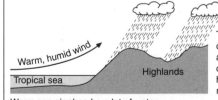

Warm air rises over cold air; it expands and cools, condensation, clouds and rain form

Cumulus cloud

This line represents the plane separating warm air from cold air

Warm air   Rain   Cold air

Warm air is forced to rise when it is undercut by colder air; clouds and sometimes rain occur

Orographic rainfall

Upland areas that encourage warm, moist air to rise; **steady rain, drizzle.**

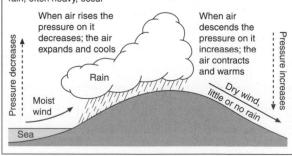

The highlands cause the humid air to rise – it cools and very heavy orographic rain falls

Warm, humid wind

Tropical sea   Highlands

Warm sea air absorbs a lot of water

When moist air is forced to rise over a mountain range, clouds and rain, often heavy, occur

When air rises the pressure on it decreases; the air expands and cools

When air descends the pressure on it increases; the air contracts and warms

Pressure decreases

Pressure increases

Rain

Moist wind

Dry wind, little or no rain

Sea

Precipitation patterns

Zone 5 Sub-tropical highs
- scanty winter rainfall
- dominated by sub-tropical highs, but occasional mid-latitude depressions bring rain

Zone 6 Mediterranean zone
- semi-arid/sub-humid region
- long dry summer, short wet winter
- sub-tropical highs in summer
- mid-latitude depressions in winter

Zone 7 Middle and high latitudes
- depressions and fronts
- precipitation in all seasons
- maximum precipitation in winter (cyclonic activity)

Zone 8 Polar regions
- low precipitation
- cold subsiding air
- some depressions in winter

THE ZONAL MODEL
- Abundance of rain in the equatorial zone; moderate to large amounts in the mid-latitudes; relatively low rainfall in the sub-tropics and particularly at the poles.
- Rainfall is abundant in the uplift areas of the convergence/convection zone of the equatorial trough (ITCZ) and in the polar frontal zones of the mid-latitudes.

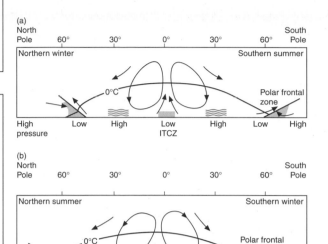

Zone 1 Equatorial zone
- abundant rainfall throughout the year associated with the permanence of the ITCZ

Zone 2 Wet and dry tropics
- wet in summer, dry in winter
- summer rains due to ITCZ
- winter dry period due to sub-tropical anticyclones

Zone 3 Tropical semi-arid
- small amount of rain in summer, very dry in other seasons
- associated with equatorial margin of sub-tropical high

Zone 4 Arid zones
- permanent dryness
- year-long dominance of sub-tropical highs

MODIFICATIONS TO THE ZONAL MODEL

Orographic barriers
- zone of mid-latitude depressions; rise due to western Cordillera
- heavy orographic precipitation along windward sides, e.g. Olympic mountains receive 3750 mm, the leeward side only 750 mm

Ocean currents
- warm air passes over cold ocean current
- the result is temperature inversion and fog on the coast
- the stability prevents convection cell
- the result is the Atacama desert

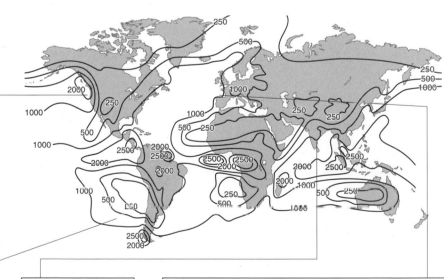

Monsoon
- shift of ITCZ gives rise to intense but strongly seasonal rainfall, e.g. India

Mid-latitude cyclonic belt
- mid-latitude depressions moving west to east
- moist air due to evaporation over the warm Atlantic
- rain dropped on western sides of continents, e.g. the UK

Mid-latitude weather systems

Mid-latitude climates occur between the tropical and polar climates in both the northern and southern hemispheres.

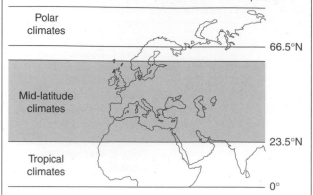

They are influenced by the meeting of warm air from the south and cold air from the north to form *fronts*. These give rise to *low pressure* systems (cyclones or depressions) and high pressure systems (anticyclones).

WHAT IS A DEPRESSION?

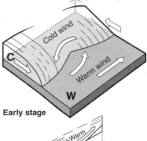

Early stage

Polar front boundary between cold and warm air.

- An instability occurs on the polar front.

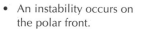
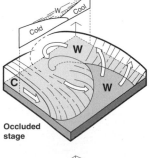

Open stage

- Cold air pushes warm air north in the northern hemisphere. Warm air rises over cold air. The boundary between the two forms the *warm front*. The colder advancing air to the west is denser and undercuts the warmer air. The boundary forms the *cold front*.

Occluded stage

- The cold front moves faster than the warm front and eventually catches it up and lifts it away from the ground, forming an *occluded* front.

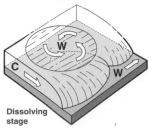

Dissolving stage

After occlusion the depression dissolves.

WHAT IS AN AIR MASS?

An area of air which has similar properties of temperature and humidity is called an air mass.

Air masses develop over areas of similar geographical character, like the polar ice caps or hot deserts.

WHAT IS A FRONT?

Fronts occur when two air masses with different temperatures and densities meet. Cold air undercuts warm air.

Temperature characteristics of a frontal zone

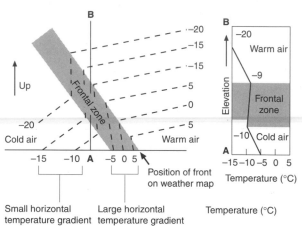

The interaction between air masses results in the formation of depressions and anticyclones.

WHAT IS AN ANTICYCLONE?

- Anticyclones develop in regions of descending air.

- Air moving towards the pole from equatorial regions descends forming sub-tropical high pressures.

- The highest pressure is in the centre.

- Winds blow outwards from the centre in a clockwise direction in the northern hemisphere (due to the Coriolis effect).

- An anticyclone is a uniform air mass which gives fair weather, especially in summer.

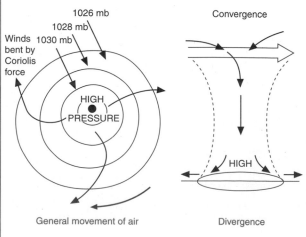

Weather associated with a depression

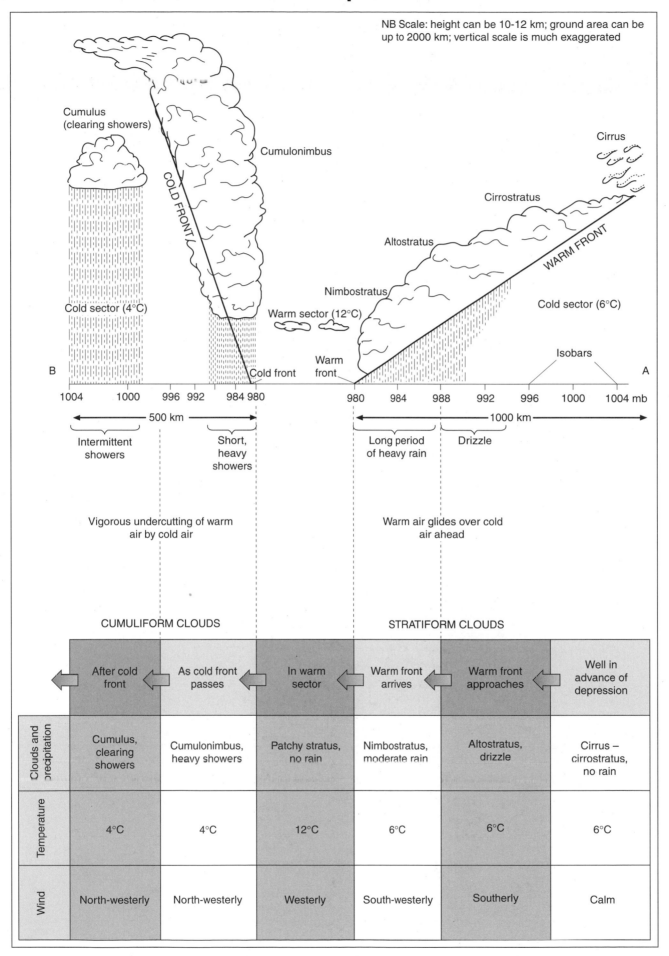

NB Scale: height can be 10-12 km; ground area can be up to 2000 km; vertical scale is much exaggerated

Cumulus (clearing showers)

Cumulonimbus

Cirrus

Cirrostratus

COLD FRONT

Altostratus

WARM FRONT

Nimbostratus

Cold sector (4°C)

Warm sector (12°C)

Cold sector (6°C)

Isobars

B   Cold front   Warm front   A

1004   1000   996   992   984   980   980   984   988   992   996   1000   1004 mb

← 500 km →   ← 1000 km →

Intermittent showers   Short, heavy showers   Long period of heavy rain   Drizzle

Vigorous undercutting of warm air by cold air

Warm air glides over cold air ahead

CUMULIFORM CLOUDS   STRATIFORM CLOUDS

	After cold front	As cold front passes	In warm sector	Warm front arrives	Warm front approaches	Well in advance of depression
Clouds and precipitation	Cumulus, clearing showers	Cumulonimbus, heavy showers	Patchy stratus, no rain	Nimbostratus, moderate rain	Altostratus, drizzle	Cirrus – cirrostratus, no rain
Temperature	4°C	4°C	12°C	6°C	6°C	6°C
Wind	North-westerly	North-westerly	Westerly	South-westerly	Southerly	Calm

The passage of a depression

A DEPRESSION CROSSING THE BRITISH ISLES

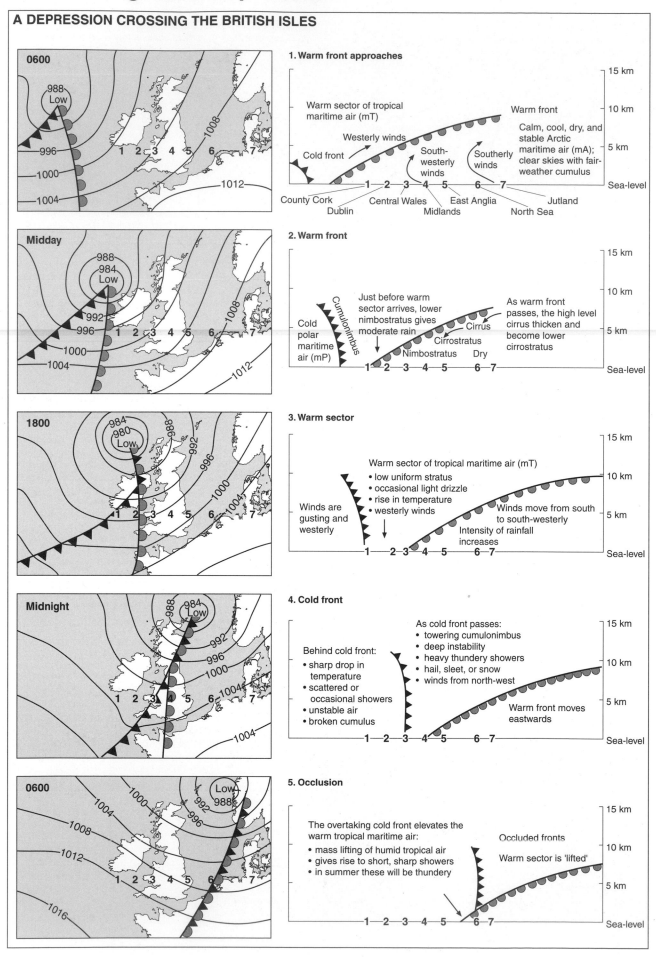

0600

Midday

1800

Midnight

0600

1. Warm front approaches

Warm sector of tropical maritime air (mT)

Westerly winds

Cold front

South-westerly winds

Warm front

Southerly winds

Calm, cool, dry, and stable Arctic maritime air (mA); clear skies with fair-weather cumulus

County Cork — Dublin — Central Wales — Midlands — East Anglia — North Sea — Jutland

2. Warm front

Cold polar maritime air (mP)

Cumulonimbus

Just before warm sector arrives, lower nimbostratus gives moderate rain

Nimbostratus

Cirrostratus

Cirrus

Dry

As warm front passes, the high level cirrus thicken and become lower cirrostratus

3. Warm sector

Warm sector of tropical maritime air (mT)
• low uniform stratus
• occasional light drizzle
• rise in temperature
• westerly winds

Winds are gusting and westerly

Winds move from south to south-westerly

Intensity of rainfall increases

4. Cold front

Behind cold front:
• sharp drop in temperature
• scattered or occasional showers
• unstable air
• broken cumulus

As cold front passes:
• towering cumulonimbus
• deep instability
• heavy thundery showers
• hail, sleet, or snow
• winds from north-west

Warm front moves eastwards

5. Occlusion

The overtaking cold front elevates the warm tropical maritime air:
• mass lifting of humid tropical air
• gives rise to short, sharp showers
• in summer these will be thundery

Occluded fronts

Warm sector is 'lifted'

Airflows and air masses

LAMB'S AIRFLOW TYPES

Lamb has identified seven major categories of airflows
(movement of pressure systems) influencing the British Isles.
Each airflow type is associated with corresponding air
masses:

Westerly	Polar maritime (mP), Tropical maritime (mT)
North-westerly	mP, Arctic maritime (mA)
Northerly	mA
Easterly	Arctic continental (cA), Polar continental (cP)
Southerly	mT or Tropical continental (cT) - summer mT or cP - winter
Cyclonic	mP, mT

Type	General weather conditions of Lamb's airflow types
Westerly	Unsettled weather with variable wind directions as depressions cross the country. Mild and stormy in winter, generally cool and cloudy in summer.
North-westerly	Cool, changeable conditions. Strong winds and showers affect windward coasts especially, but the southern part of Britain may have dry, bright weather.
Northerly	Cold weather at all seasons, often associated with polar lows or troughs. Snow and sleet showers in winter, especially in the north and east.
Easterly	Cold in the winter half-year, sometimes very severe weather in the south and east with snow or sleet. Warm in summer with dry weather in the west. Occasionally thundery.
Southerly	Generally warm and thundery in summer. In winter it may be associated with a depression in the Atlantic, giving mild, damp weather, especially in the south-west, or with a high over central Europe, in which case it is cold and dry.
Cyclonic	Rainy, unsettled conditions often accompanied by gales and thunderstorms. This type may refer either to the rapid passage of depressions across the country or to the persistence of a deep depression.
Anticyclonic	Warm and dry in summer apart from occasional thunderstorms. Cold in winter with night frosts and fog, especially in autumn.

AVERAGE AIR-MASS FREQUENCIES FOR KEW (LONDON)

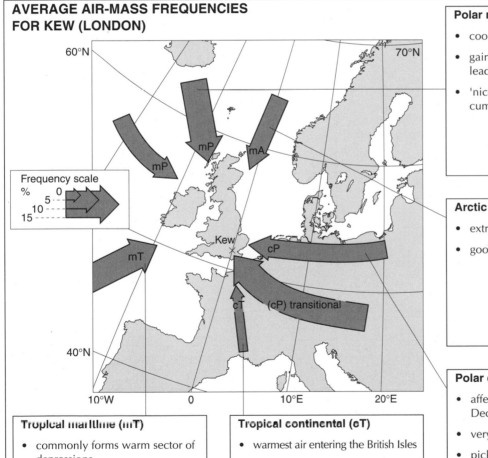

Polar maritime (mP)

- cool, showery weather
- gains moisture over the sea, leading to unstable air
- 'nice morning, bad day' cumuliform clouds

Arctic maritime (mA)

- extreme weather
- good visibility

Polar continental (cP)

- affects British Isles between December and February
- very cold, dry air from Siberia
- picks up moisture from the sea and can lead to snow showers, especially on the east coast
- wind chill factor (dry air) exaggerates coldness

Tropical maritime (mT)

- commonly forms warm sector of depressions
- in winter, air is unseasonally mild and damp
- stratos or stratocumulus cloud with drizzle

Tropical continental (cT)

- warmest air entering the British Isles
- can lead to heatwaves or late summer warming – the September 'Indian summer'
- can lead to instability and thunderstorms
- in winter it can bring fine, hazy, mild weather

Low-latitude (tropical) weather systems

The climate of the tropics is associated with variations in rainfall and dramatic phenomena like tropical disturbances, hurricanes, and the monsoon.

THE ATMOSPHERE IN THE TROPICS

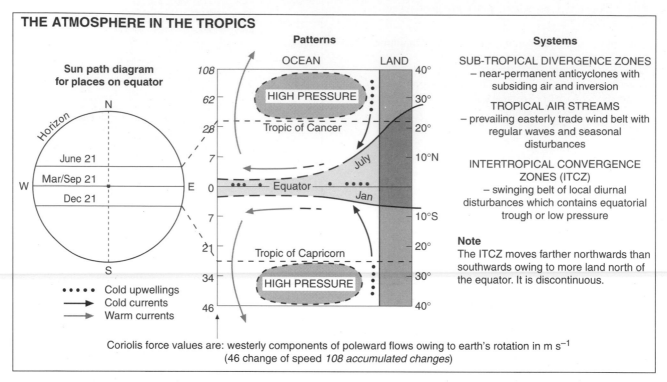

Sun path diagram for places on equator

Cold upwellings
→ Cold currents
→ Warm currents

Patterns

OCEAN   LAND

HIGH PRESSURE

Tropic of Cancer

July

Equator

Jan

Tropic of Capricorn

HIGH PRESSURE

Coriolis force values are: westerly components of poleward flows owing to earth's rotation in m s^{-1}
(46 change of speed *108 accumulated changes*)

Systems

SUB-TROPICAL DIVERGENCE ZONES
– near-permanent anticyclones with subsiding air and inversion

TROPICAL AIR STREAMS
– prevailing easterly trade wind belt with regular waves and seasonal disturbances

INTERTROPICAL CONVERGENCE ZONES (ITCZ)
– swinging belt of local diurnal disturbances which contains equatorial trough or low pressure

Note
The ITCZ moves farther northwards than southwards owing to more land north of the equator. It is discontinuous.

TROPICAL DISTURBANCES

Some storms are associated with wave disturbances. They have a wavelength of about 3000 km from east to west. The passage of waves gives rise to storms.

Direction of movement

RIDGE

Cloud cover

DIVERGENCE

CONVERGENCE

TROUGH

H

X
Y

0   km   600

km
15
10
5
0
Height

X 400 200 0 200 400 600 Y

← Surface stream-lines
◄--- 200 mb stream-lines
•••• Top of moist layer

Fine weather, scattered cumulus cloud, some haze

Well-developed cumulus, occasional showers, improving visibility

Veer of wind direction, heavy cumulus and cumulonimbus, moderate or heavy thundery showers, and a decrease of temperature

HURRICANES (CYCLONIC DISTURBANCES)

Hurricanes originate over warm oceans with sea temperatures in excess of 27°C, with air blowing inwards and flowing out at upper levels.

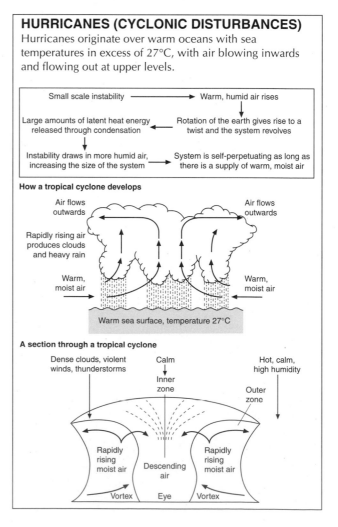

Small scale instability ——→ Warm, humid air rises

Large amounts of latent heat energy released through condensation ←— Rotation of the earth gives rise to a twist and the system revolves

Instability draws in more humid air, increasing the size of the system ——→ System is self-perpetuating as long as there is a supply of warm, moist air

How a tropical cyclone develops

Air flows outwards

Air flows outwards

Rapidly rising air produces clouds and heavy rain

Warm, moist air

Warm, moist air

Warm sea surface, temperature 27°C

A section through a tropical cyclone

Dense clouds, violent winds, thunderstorms

Calm

Hot, calm, high humidity

Inner zone

Outer zone

Rapidly rising moist air

Descending air

Rapidly rising moist air

Vortex   Eye   Vortex

The monsoon

WHAT IS THE MONSOON?

The monsoon is the reversal of pressure and winds which gives rise to a marked seasonality of rainfall over south and south-east Asia.

This can be seen clearly in India and Bangladesh.

A number of influences have been suggested:
- the effect of the Himalayas on the ITCZ
- reduced CO_2, due to the Tibetan Plateau, leads to cooling in the interior
- differential heating and cooling of land compared to the adjacent seas

WINTER: NORTH-EAST MONSOON

- Temperatures over central Asia are low, leading to high pressure.

- Jet stream splits into two; the southern sub-tropical jet leads to descending air and high pressure.

- This leads outward-blowing north-easterly winds across south Asia.
- These dry airstreams produce clear skies and sunny weather over most of India (November–May).

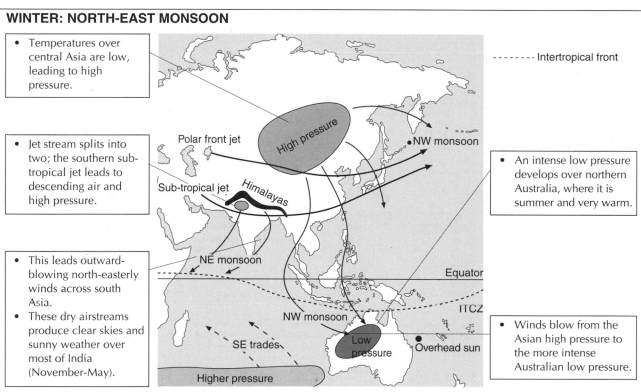

- An intense low pressure develops over northern Australia, where it is summer and very warm.

- Winds blow from the Asian high pressure to the more intense Australian low pressure.

SUMMER: SOUTH-WEST MONSOON

- In March and May the winds shift, and the upper westerly air currents begin to move north.
- The jet stream strengthens until it lies entirely to the north of the Himalayas.

- The overhead sun migrates north to a position just over India and the ITCZ moves north (the monsoon trough).

- Intense low pressure develops over Asia separated, by the Himalayas, from a smaller intense low pressure over the Punjab.

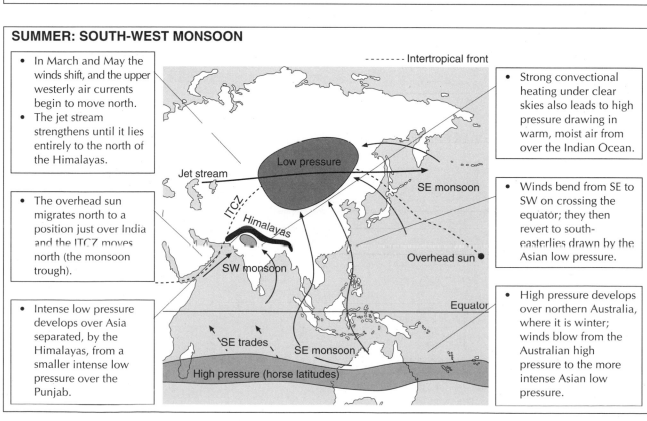

- Strong convectional heating under clear skies also leads to high pressure drawing in warm, moist air from over the Indian Ocean.

- Winds bend from SE to SW on crossing the equator; they then revert to south-easterlies drawn by the Asian low pressure.

- High pressure develops over northern Australia, where it is winter; winds blow from the Australian high pressure to the more intense Asian low pressure.

Local winds

MOUNTAIN AND VALLEY WINDS

Anabatic winds (up-valley) form during warm afternoons due to greater heating of valley sides compared with the valley floor.

Katabatic winds (down-valley) reverse the process as the cold, denser air at higher elevations drains into depressions and valleys.

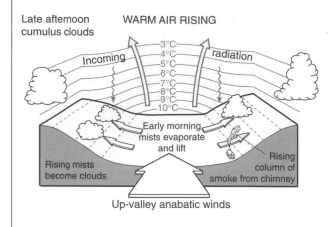

Late afternoon cumulus clouds
WARM AIR RISING
Incoming radiation
3°C
4°C
5°C
6°C
7°C
8°C
9°C
10°C
Early morning mists evaporate and lift
Rising mists become clouds
Rising column of smoke from chimney
Up-valley anabatic winds

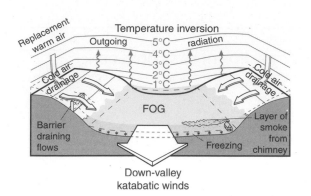

Replacement warm air
Temperature inversion
Outgoing radiation
5°C
4°C
3°C
2°C
1°C
Cold air drainage
Cold air drainage
Barrier draining flows
FOG
Freezing
Layer of smoke from chimney
Down-valley katabatic winds

LAND AND SEA BREEZES

A **daytime sea breeze** is an *onshore wind* which occurs because land temperatures rise more rapidly during the day.

A **nocturnal land breeze** is an *offshore wind* generated by the sea cooling less rapidly than the land at night (also downslope winds blowing off the land).

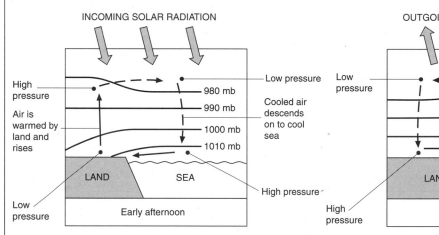

INCOMING SOLAR RADIATION
High pressure
Low pressure
980 mb
990 mb
1000 mb
1010 mb
Air is warmed by land and rises
Cooled air descends on to cool sea
LAND
SEA
Low pressure
High pressure
Early afternoon

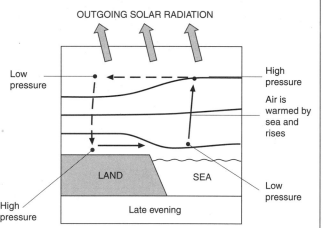

OUTGOING SOLAR RADIATION
Low pressure
High pressure
Air is warmed by sea and rises
LAND
SEA
High pressure
Low pressure
Late evening

FOHN WINDS

Flows of air over mountains can create a **fohn effect**. This can result in a rapid increase in temperature and falls in relative humidity. The winds melt snow, leading to avalanches and flooding. The low humidity fohn wind can also dry forest areas, causing fires.

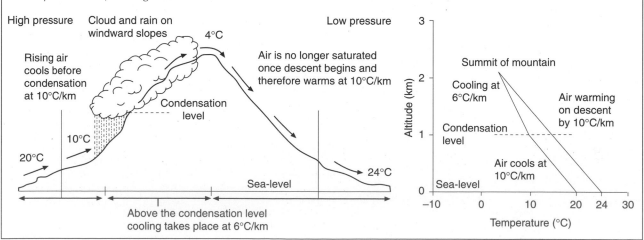

High pressure
Cloud and rain on windward slopes
4°C
Low pressure
Rising air cools before condensation at 10°C/km
Condensation level
Air is no longer saturated once descent begins and therefore warms at 10°C/km
10°C
20°C
Sea-level
24°C
Above the condensation level cooling takes place at 6°C/km

Altitude (km)
3
2
1
0
Summit of mountain
Cooling at 6°C/km
Air warming on descent by 10°C/km
Condensation level
Air cools at 10°C/km
Sea-level
Temperature (°C)
−10 0 10 20 24 30

Urban microclimates

Structure of the air above the urban area

Greater amounts of dust mean increasing concentrations of hygroscopic particles. Less water vapour, but more CO_2 and higher proportions of noxious fumes owing to combustion of imported fuels. Discharge of waste gases by industry.

Structure of the urban surface

More heat-retaining materials with lower albedo and better radiation absorbing properties. Rougher surfaces with a great variety of perpendicular slopes facing different aspects. Tall buildings can be very exposed, and the deep streets are sheltered and shaded.

Resultant processes

1. Radiation and sunshine

Greater scattering of shorter-wave radiation by dust, but much higher absorption of longer waves owing to surfaces and CO_2. Hence more diffuse sky radiation with considerable local contrasts owing to variable screening by tall buildings in shaded narrow streets. Reduced visibility arising from industrial haze.

2. Clouds and fogs

Higher incidence of thicker cloud covers in summer and radiation fogs or smogs in winter because of increased convection and air pollution respectively. Concentrations of hygroscopic particles accelerate the onset of condensation (see 5 below).

3. Temperatures

Stronger heat energy retention and release, including fuel combustion, gives significant temperature increases from suburbs into the centre of built-up areas creating heat 'islands'. These can be up to 8°C warmer during winter nights. Heating from below increases air mass instability overhead, notably during summer afternoons and evenings. Big local contrasts between sunny and shaded surfaces, especially in the spring.

4. Pressure and winds

Severe gusting and turbulence around tall buildings causing strong local pressure gradients from windward to leeward walls. Deep narrow streets much calmer unless aligned with prevailing winds to funnel flows along them.

5. Humidity

Decreases in relative humidity occur in inner cities owing to lack of available moisture and higher temperatures there. Partly countered in very cold stable conditions by early onset of condensation in low-lying districts and industrial zones (see 2 above.).

6. Precipitation

Perceptibly more intense storms, particularly during hot summer evenings and nights owing to greater instability and stronger convection above built-up areas. Probably higher incidence of thunder in appropriate locations. Less snowfall and briefer covers even when uncleared.

The effect of city morphology on radiation received at the surface

(a) Isolated buildings
Isolated building
Sunny side heated by insolation, reflected insolation, radiation, and conduction
Heat stored and re-radiated
Shaded side

(c) High buildings
Very little radiation reaches street level. Radiation reflected off lower walls after reflection from near tops of buildings

(b) Low buildings
Street collects reflected radiation

The structure of the urban climatic dome

Prevailing wind
Urban boundary layer
Urban plume develops downwind
Urban canopy layer below roof level
Rural boundary layer
RURAL   SUBURBAN   URBAN   SUBURBAN   RURAL

The morphology of the urban heat island

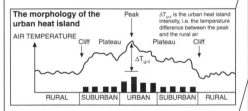

Peak
ΔT_{u-r} is the urban heat island intensity, i.e. the temperature difference between the peak and the rural air
AIR TEMPERATURE
Cliff   Plateau   Plateau   Cliff
ΔT_{u-r}
RURAL   SUBURBAN   URBAN   SUBURBAN   RURAL

Airflow modified by a single building

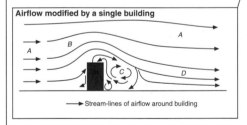

A
B
A
C
D
→ Stream-lines of airflow around building

THE URBAN HEAT ISLAND

Urban areas are generally warmer than those of the surrounding countryside. Temperatures are on average 2-4°C higher in urban areas. This creates an **urban heat island**. It can be explained by heat and pollution release.

Lower wind speeds due to the height of buildings and urban surface roughness.	Urban pollution and photochemical smog can trap outgoing radiant energy.	Burning of fossil fuels for domestic and commercial use can exceed energy inputs from the sun.
Buildings have a higher capacity to retain and conduct heat and a lower albedo.	Reduction in thermal energy required for evaporation and evapotrans-piration due to the surface character, rapid drainage, and generally lower wind speeds.	Reduction of heat diffusion due to changes in airflow patterns as the result of urban surface roughness.

Condensation and lapse rates

CONDENSATION LEVEL

A rising, expanding parcel of air will cool until it reaches **dew-point** temperature. At this point there is insufficient energy to keep water as a vapour and condensation occurs, i.e. actual vapour pressure = saturated vapour pressure and relative humidity = 100%.

SATURATED VAPOUR PRESSURE

This is the maximum moisture vapour that air can hold at a given temperature. As temperature decreases, air can hold less vapour and so condensation occurs.

$$\text{Relative humidity} = \frac{\text{actual vapour pressure}}{\text{saturated vapour pressure}} \times 100$$

WHAT ARE THE CONDITIONS FOR CONDENSATION?

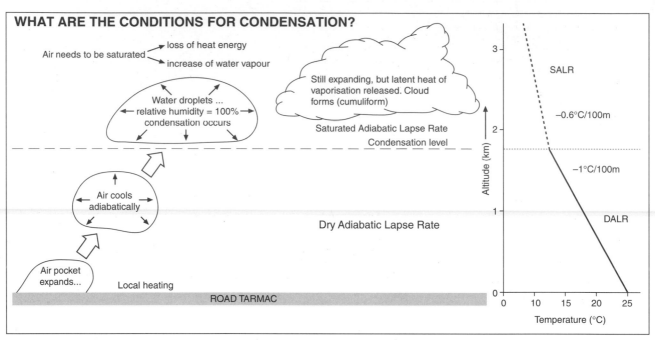

LAPSE RATES

Stable air
- air rises until surrounding air temperature is warmer than the rising pocket
- rising air cools and sinks back

Unstable air
- air continues to rise when lifting stimulus (ground heating) is relaxed
- vertical currents dominate
- cumulonimbus clouds and thunderstorms

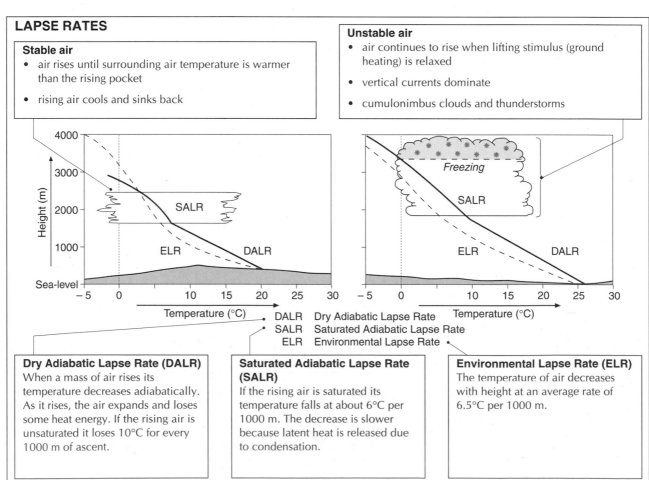

DALR   Dry Adiabatic Lapse Rate
SALR   Saturated Adiabatic Lapse Rate
ELR   Environmental Lapse Rate

Dry Adiabatic Lapse Rate (DALR)
When a mass of air rises its temperature decreases adiabatically. As it rises, the air expands and loses some heat energy. If the rising air is unsaturated it loses 10°C for every 1000 m of ascent.

Saturated Adiabatic Lapse Rate (SALR)
If the rising air is saturated its temperature falls at about 6°C per 1000 m. The decrease is slower because latent heat is released due to condensation.

Environmental Lapse Rate (ELR)
The temperature of air decreases with height at an average rate of 6.5°C per 1000 m.

Soils

Soils form the outer-most part of the earth's surface, and are made up of weathered bedrock (regolith), organic matter, air, and water. Soil has matter in all three states - solid, liquid, and gaseous.

SOIL TEXTURE

Soil texture refers to the size of the solid particles in a soil, ranging from gravel to clay. The proportions vary from soil to soil and between horizons within a soil.

(Soil texture triangle diagram with axes Clay (%), Silt (%), and Sand (%). Regions labelled: Clay, Silty clay, Sandy clay, Clay loam, Silty clay loam, Sandy clay loam, Loam, Sandy clay, Silty loam, Loamy sand, Sand, Silty.)

Soil texture is important as it affects:
- moisture content and aeration
- retention of nutrients
- ease of cultivation and root penetration

In general, clay soils become water-logged whereas sandy soils drain rapidly. A **loam** (mixed soil) is best for plants.

Particle size - diameter (mm)

Clay	<0.002
Silt	<0.02
Sand	<2
Gravel	≥2

SOIL STRUCTURE

The shape of the soil clumps (or aggregates) is known as structure.

Crumb	1-6 mm
Platy	1-10 mm
Block	5-75 mm
Prismatic	10-100 mm
Columnar	10-100 mm

SOIL MOISTURE

Saturation

- all pore spaces filled with water
- some water drains as a result of gravity

Field capacity

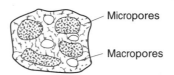

Micropores

Macropores

- small micropores filled with water and held by suction
- macropores (large pores) filled with air
- water available to plants

Wilting

- water present only in small quantities, held by soil hygroscopically

SOIL CHEMISTRY

Nutrients in the soil are called bases. Plants use bases for growth and in return provide the soil with hydrogen ions. Consequently, soils become more acidic over time (acidity refers to the proportion of exchangeable hydrogen ions present). However, bases can be returned via leaf fall, application of fertilisers, or the weathering of soft base-rich rocks such as chalk. The **cation exchange capacity** is the ability to retain positively-charged ions. A soil with a high CEC is more fertile than a soil with a low CEC.

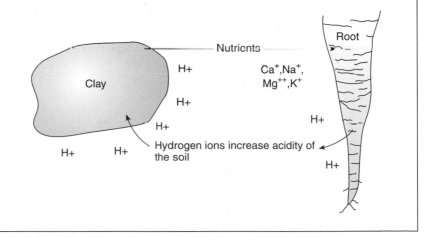

Soil formation

Soil formation is affected by a number of factors, most notably climate, geology, biological organisms, and topography. These interact over time to produce distinctive soils and soil profiles.

CLIMATE

Two important **climate** mechanisms exist:
- precipitation effectiveness
- temperature

Precipitation effectiveness is a measure of the extent to which precipitation (Ppt) exceeds potential evapotranspiration (P.Evt).
- If Ppt > P.Evt, there is a downward movement of materials in the soil, and the soil is leached.
- If Ppt < P.Evt, there is an upward movement of material by capillary action.

Temperature affects the rate of biological and chemical action. In general, as temperatures increase so does chemical and biological activity.

The **zonal** classification of soil states that on a global scale soils are determined by climate.

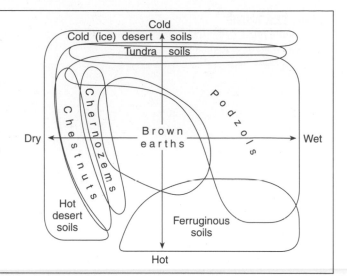

GEOLOGY

Geology has a lasting effect on soils through **texture**, **structure**, and **fertility**. Sandstones produce free-draining soils whereas clays and shales give much finer soils. The distinction between calcareous rocks, e.g. chalk, which are base-rich or basic, and non-calcareous rocks, e.g. granite, which are base-poor or acidic, is important. On a regional scale soils often vary with geology, as in the case of the Isle of Purbeck.

The **intrazonal** classification states that within a climatic zone soils vary with rock type.

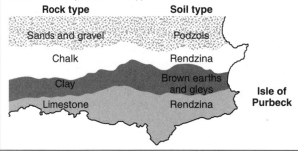

Rock type	Soil type
Sands and gravel	Podzols
Chalk	Rendzina
Clay	Brown earths and gleys
Limestone	Rendzina

Isle of Purbeck

BIOTIC FACTORS

Biotic factors include micro-organisms breaking down leaf litter, worms mixing soils, vegetation returning nutrients, and people adding fertiliser and irrigating, draining, and compacting soils.

TIME

Time is not a causative factor, but allows processes to operate. The time needed for soil formation varies. Sandstones develop soils more quickly than granite or basalt. Some British soils have evolved since the last glaciation. Soils which have not had enough time to properly mature are termed **azonal** soils.

TOPOGRAPHY AND CATENAS

Altitude	1000 m
Rainfall	1000 mm
Average temperature	7°C

- Steeper slopes have thinner soils.
- Soil erosion increases with slope angle.
- Aspect affects microclimate.
- A catena is the variation in soils along a slope owing to changes in slope angle and drainage, climate, and water table. Rock type is constant.

Altitude	250 m
Rainfall	750 mm
Average temperature	11°C

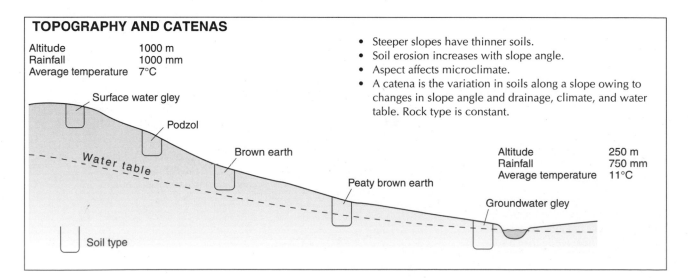

Soil-forming processes

SOIL HORIZONS

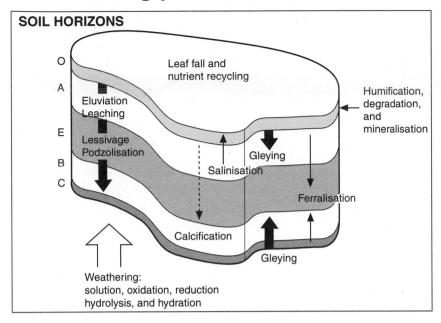

Leaf fall and nutrient recycling

O
A
Eluviation
Leaching
E
Lessivage
Podzolisation
B
C

Gleying
Salinisation
Ferralisation
Calcification
Gleying

Humification, degradation, and mineralisation

Weathering:
solution, oxidation, reduction hydrolysis, and hydration

Humification, **degradation**, and **mineralisation** are the processes whereby organic matter is broken down and the nutrients are returned to the soil. The breakdown releases organic acids, **chelating agents**, which break down clay to silica and soluble iron and aluminium.

Cheluviation is the removal of the iron and aluminium sesquioxides under the influence of chelating agents.

Illuviation is the redeposition of material in the lower horizons.

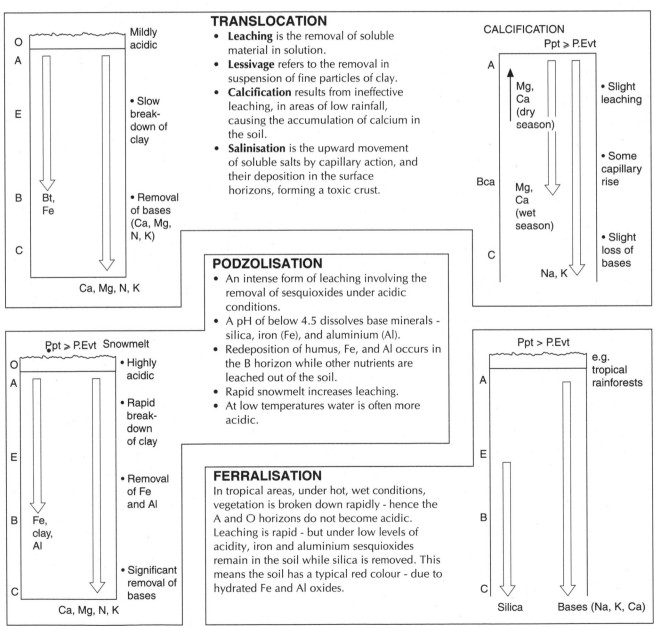

TRANSLOCATION

- **Leaching** is the removal of soluble material in solution.
- **Lessivage** refers to the removal in suspension of fine particles of clay.
- **Calcification** results from ineffective leaching, in areas of low rainfall, causing the accumulation of calcium in the soil.
- **Salinisation** is the upward movement of soluble salts by capillary action, and their deposition in the surface horizons, forming a toxic crust.

O
A
E
B Bt, Fe
C

Mildly acidic

- Slow breakdown of clay
- Removal of bases (Ca, Mg, N, K)

Ca, Mg, N, K

CALCIFICATION
Ppt ≥ P.Evt

A
Mg, Ca (dry season)
Bca
Mg, Ca (wet season)
C
Na, K

- Slight leaching
- Some capillary rise
- Slight loss of bases

PODZOLISATION

- An intense form of leaching involving the removal of sesquioxides under acidic conditions.
- A pH of below 4.5 dissolves base minerals - silica, iron (Fe), and aluminium (Al).
- Redeposition of humus, Fe, and Al occurs in the B horizon while other nutrients are leached out of the soil.
- Rapid snowmelt increases leaching.
- At low temperatures water is often more acidic.

Ppt ≥ P.Evt Snowmelt

O
A
E
B Fe, clay, Al
C

- Highly acidic
- Rapid breakdown of clay
- Removal of Fe and Al
- Significant removal of bases

Ca, Mg, N, K

FERRALISATION

In tropical areas, under hot, wet conditions, vegetation is broken down rapidly - hence the A and O horizons do not become acidic. Leaching is rapid - but under low levels of acidity, iron and aluminium sesquioxides remain in the soil while silica is removed. This means the soil has a typical red colour - due to hydrated Fe and Al oxides.

Ppt > P.Evt

A
E
B
C

e.g. tropical rainforests

Silica   Bases (Na, K, Ca)

Soil types

SPODOSOLS	Podzols
ALFISOLS	Brown earths (brunizems)
MOLLISOLS	Chernozems
ARIDISOLS	Desert and semi-desert
ULTISOLS	Ferruginous soils of the savanna (pronounced wet and dry seasons)
OXISOLS	Ferralitic soils of the tropical rainforest (latosols)

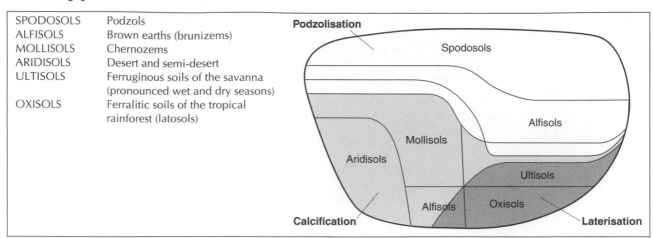

Podzols

Coniferous/heathland vegetation, mor humus, pH 4.5 or less

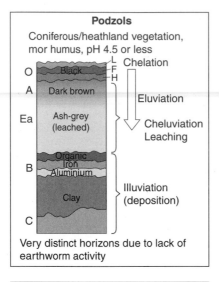

Very distinct horizons due to lack of earthworm activity

Brown earth

Deciduous vegetation, mull humus, mildly acidic, pH 5.5-6.5

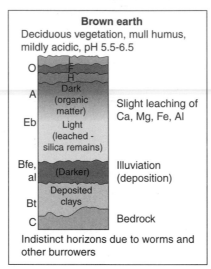

Indistinct horizons due to worms and other burrowers

Chernozem

Grassland vegetation, Ppt = P.Evt

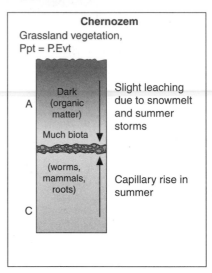

Ferralitic (latosol)

Tropical rainforest, Ppt > P.Evt

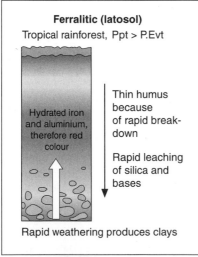

Rapid weathering produces clays

Groundwater gley

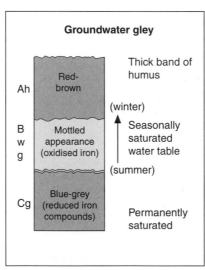

Rendzina

Grass vegetation

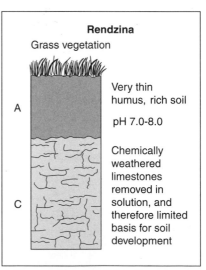

Surface water gley

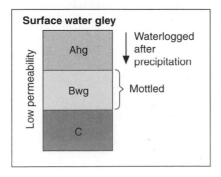

Ferruginous soil

Savanna vegetation, distinct wet and dry seasons

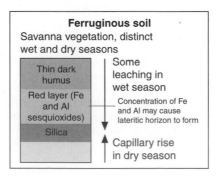

Saline soil

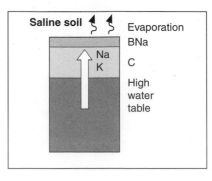

Human impact on soils

SOIL EROSION

Soil erosion in North America: the effect of land use

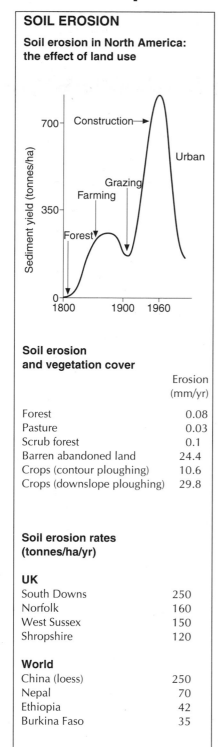

Soil erosion and vegetation cover

	Erosion (mm/yr)
Forest	0.08
Pasture	0.03
Scrub forest	0.1
Barren abandoned land	24.4
Crops (contour ploughing)	10.6
Crops (downslope ploughing)	29.8

Soil erosion rates (tonnes/ha/yr)

UK

South Downs	250
Norfolk	160
West Sussex	150
Shropshire	120

World

China (loess)	250
Nepal	70
Ethiopia	42
Burkina Faso	35

INTENSIVE AGRICULTURE AND ITS EFFECT ON SOIL PROFILES

Soil characteristic	Natural state	Intensive agriculture
Organic content	A horizon high (7%) B horizon 0%	Uniform (3-5%) in ploughed horizon
Carbonates	A horizon low/zero B/C horizon maximum	Uniform if limed and tilled
Nitrogen	Medium/low	High (nitrate fertiliser)
Biological activity	High	Medium
Exchangeable cation balance	Ca 80% K 5% P 3% H 7%	Ca 80% K 5% P 3% H 7%

SOIL COMPACTION

A change in structure can occur, turning a free-draining soil or horizon into a compact, impermeable soil or horizon. A **plough pan** forms, caused when damp soil is ploughed and moulded. Compaction by the weight of modern machinery increases the problem.

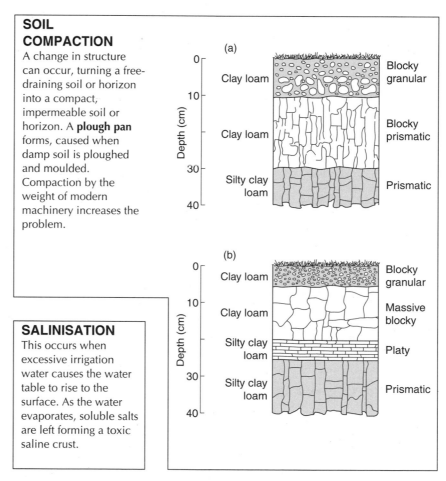

SALINISATION

This occurs when excessive irrigation water causes the water table to rise to the surface. As the water evaporates, soluble salts are left forming a toxic saline crust.

PRINCIPLES OF GOOD SOIL MANAGEMENT

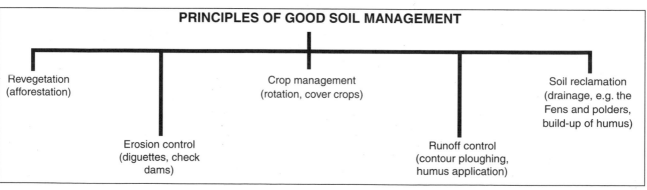

Revegetation (afforestation)

Erosion control (diguettes, check dams)

Crop management (rotation, cover crops)

Runoff control (contour ploughing, humus application)

Soil reclamation (drainage, e.g. the Fens and polders, build-up of humus)

Ecosystems

An **ecosystem** is the interrelationship between plants, animals, and their living and non-living environments. **Biogeography** is the geographic distribution of soils, vegetation, and ecosystems - where they are and why they are there.

Ecosystems can be divided into two main components:

- **Abiotic** elements (non-living), e.g. air, water, heat, nutrients, rock, and sediments.

- **Biotic** elements (living), e.g. plants and animals. These can be divided into:
  Autotrophs (or producers) - organisms capable of converting sunlit energy into food energy by photosynthesis.
  Heterotrophs (or consumers) - organisms that must feed on other organisms, e.g.
  herbivores - plant eaters
  carnivores - meat eaters
  omnivores - plant and meat eaters
  detritivores - decomposers

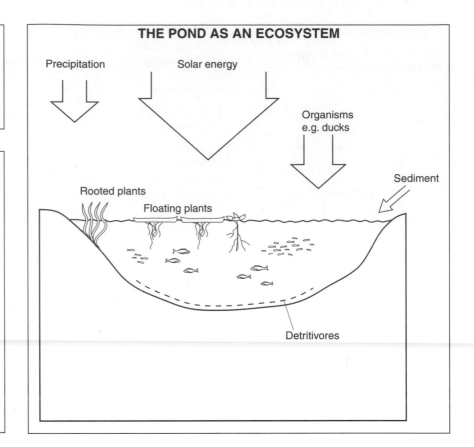

THE POND AS AN ECOSYSTEM

Precipitation Solar energy Organisms e.g. ducks

Sediment

Rooted plants

Floating plants

Detritivores

The **trophic classification** or system is based on feeding patterns. Typically there is a **trophic pyramid**, showing a larger plant biomass and a smaller consumer biomass. This occurs because:

(i) no energy transfer is 100% efficient - the transfer of light energy to food energy is only 1% efficient;

(ii) there are large losses of energy at each trophic level due to respiration, growth, reproduction, mobility, and so on.

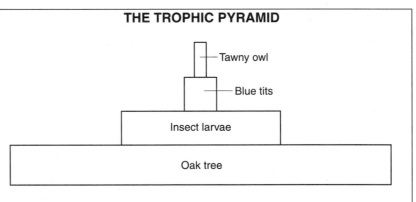

THE TROPHIC PYRAMID

Tawny owl

Blue tits

Insect larvae

Oak tree

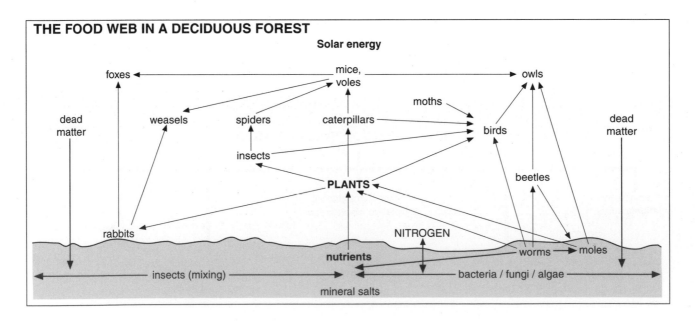

THE FOOD WEB IN A DECIDUOUS FOREST

Solar energy

foxes mice, voles owls

dead matter

weasels spiders caterpillars moths

birds

insects

dead matter

beetles

PLANTS

rabbits

NITROGEN

worms moles

nutrients

insects (mixing) bacteria / fungi / algae

mineral salts

Ecosystems and nutrient cycles

Productivity refers to the rate of energy production, normally on an annual basis.

Primary productivity refers to plant productivity.

Secondary productivity refers to that produced by animals.

Gross productivity is the total amount of energy fixed.

Net productivity is the amount of energy left after losses to respiration, growth, and so on, are taken into account.

Net primary productivity (NPP) is the amount of energy made available by plants to animals at the herbivore level. It is normally expressed as $g/m^2/yr$. NPP depends upon the amount of **heat**, **moisture**, **nutrient availability**, and **competition**, the number of **sunlight hours**, the **age of plants**, and the **health of plants**. In geographic terms NPP increases towards the equator, water permitting, and declines towards the poles.

Ecosystem	Mean NPP ($kg/m^2/yr$)	Mean biomass (kg/m^2)
Tropical rainforest	2.2	45
Tropical deciduous forest	1.6	35
Tropical scrub	0.37	3
Savanna	0.9	4
Mediterranean sclerophyll	0.5	6
Desert	0.003	0.002
Temperate grassland	0.6	1.6
Temperate forest	1.2	32.5
Boreal forest	1.2	32.5
Tundra and mountain	0.14	0.6
Open ocean	0.12	0.003
Continental shelf	0.36	0.001
Estuaries	1.5	1

GERSMEHL'S NUTRIENT CYCLES

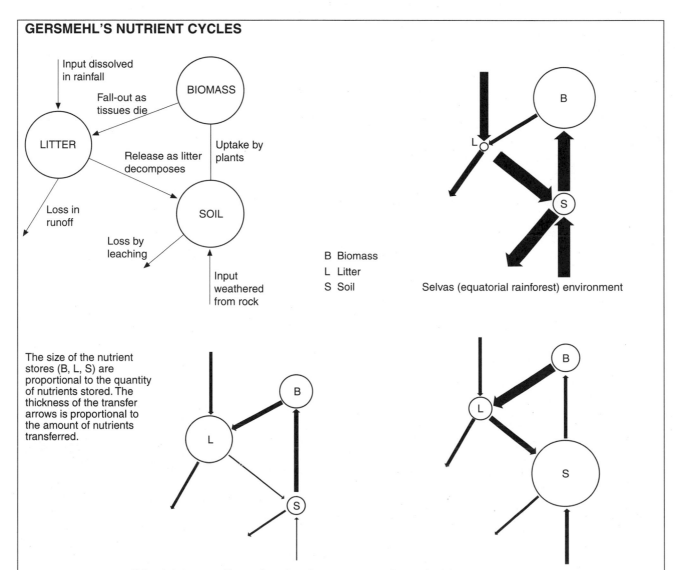

The size of the nutrient stores (B, L, S) are proportional to the quantity of nutrients stored. The thickness of the transfer arrows is proportional to the amount of nutrients transferred.

B Biomass
L Litter
S Soil

Selvas (equatorial rainforest) environment

Taiga (northern coniferous forest) environment

Steppe (mid-latitude continental grassland) environment

Nutrient cycles can be **sedimentary based**, i.e. the source of the nutrient is from rocks, or it can be **atmospheric based**, as in the case of the nitrogen cycle. Nutrient cycles can be shown by means of simplified diagrams - indicating the stores of nutrients as well as the transfers. The most important factors which determine these are availability of moisture, heat, fire (in grasslands), density of vegetation, competition, and length of growing season.

Succession

Succession, or **prisere**, is the sequential change in species in a plant community as it moves towards a **seral climax**. Each **sere** is an association or group of species, which alters the micro-environment and allows another group of species to dominate. The **climax community** is the group of species that are at a **dynamic equilibrium** with the prevailing environmental conditions - in the UK, under natural conditions, this would be oak woodland. On a global scale, climate is the most important factor in determining large scale vegetation groupings or **biomes**, e.g. rainforest, temperate grassland, and so on. However, in some areas, vegetation distribution may be influenced more by soils than climate. This is known as **edaphic** control. In savanna areas forests dominate clay soils, grassland sandy soils. Soils may also affect plant groupings on a local scale, within a climatic region, e.g. on the Isle of Purbeck grassland is found on limestone and chalk rendzina soils, forest on the brown earths, and heathland on the podzols associated with sands and gravels.

SUCCESSION AND SPECIES SELECTION

r-species are the initial colonisers - large numbers of a few species. Highly adaptable, rapid development, early reproduction, small in size, short life, highly productive.

k-species are diverse, and are specialists - a few of many species. Slower development, delayed reproduction, larger size, longer living, less productive.

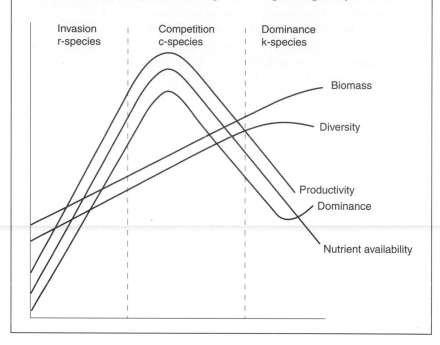

A **plagioclimax** refers to a plant community permanently influenced by people, i.e. it is prevented from reaching climatic climax by burning, grazing, and so on. Britain's **moorlands** are a good example: deforestation, burning, and grazing have replaced the original oak woodland.

A MODEL OF SUCCESSION

Rooted plants increase sedimentation and nutrient availability. They alter the micro-environment so that reeds, fen, carr, and oak are increasingly able to tolerate the thickening and drying soil.

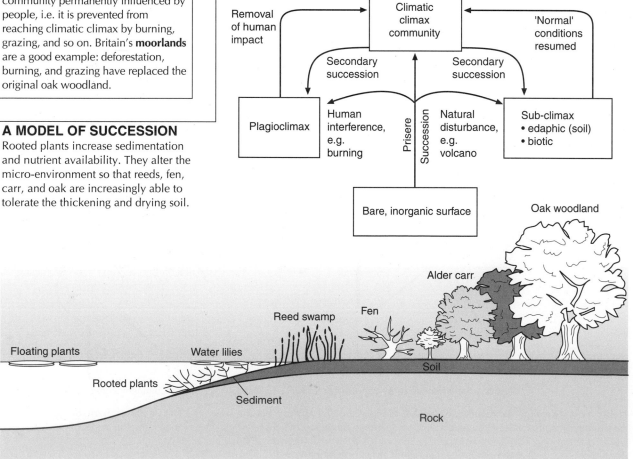

Tropical rainforests

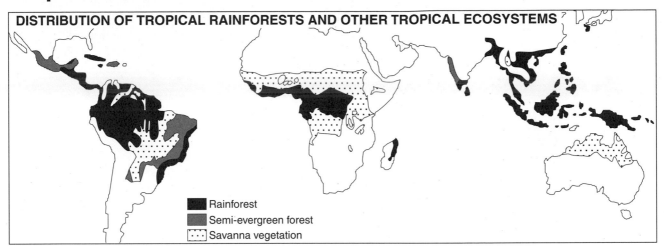

- ■ Rainforest
- ▨ Semi-evergreen forest
- ⠿ Savanna vegetation

CLIMATE

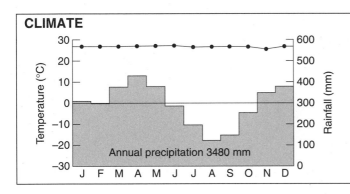

Annual precipitation 3480 mm

Some features of climates in rainforest areas:
- Annual temperatures are high (26-27°C), owing to the equatorial location of rainforest areas.
- Seasonal temperature ranges are low, 1-2°C, and diurnal (daily) ranges are greater, 10-15°C.
- Rainfall is high (>2000 mm per year), intense, convectional, and occurs on about 250 days each year.
- Humidity levels on the ground are high, often 100%.
- The growing season is year-round.

VEGETATION

- The vegetation is evergreen, enabling photosynthesis to take place year-round.
- It is layered, and the shape of the crowns vary with the layer, in order to receive light.
- Rainforests are a very productive ecosystem: NPP is about 2200 g/m²/yr, and there is a large amount of stored energy (biomass 45 kg/m).
- The ecosystem is diverse - there can be as many as 200 species of tree per hectare (an area the size of a rugby pitch), including figs, teak, mahogany, and yellow-woods.

A
Wide-spaced umbrella-shaped crowns, straight trunks, and high branches

B
Medium-spaced mop-shaped crowns

C
Densely-packed conical-shaped crowns

D
Sparse vegetation of shrubs and saplings

F Root layers

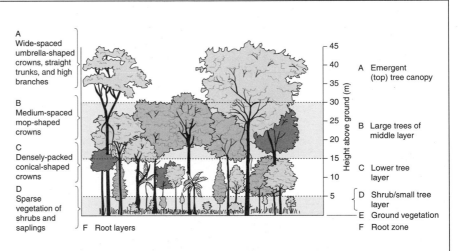

A Emergent (top) tree canopy

B Large trees of middle layer

C Lower tree layer

D Shrub/small tree layer
E Ground vegetation
F Root zone

SOILS: TROPICAL LATOSOLS

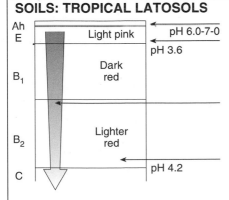

Ah E	Light pink	← pH 6.0-7.0
		pH 3.6
B₁	Dark red	
B₂	Lighter red	
C		pH 4.2

Thin layer of humus. Continuous supply of leaves: rate of humus turnover 1%/day. Very active soil fauna.

Accumulation of iron and aluminium gives the soil a red colour. Concentration of aluminium may form bauxite nodules.

Hot, wet conditions cause rapid removal of fine clay and silicate particles.

Loss of N, Ca, Mg, and K from the soil by leaching.

Bedrock intensely weathered - due to hot, wet conditions, and lack of disturbance in glacial times.

Destruction of the tropical rainforest

DEFORESTATION

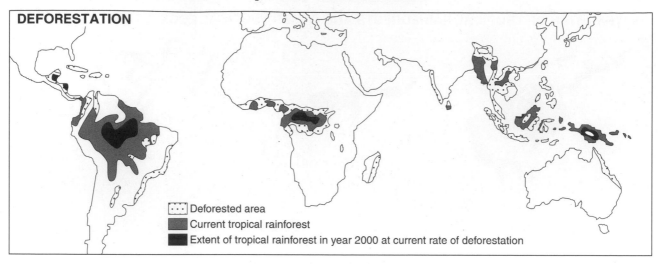

- :::: Deforested area
- ▨ Current tropical rainforest
- ■ Extent of tropical rainforest in year 2000 at current rate of deforestation

THE FRAGILE RAINFOREST ENVIRONMENT

Despite containing some of the world's most luxuriant vegetation, tropical rainforests are found on some of the world's least fertile soils. This paradox is explained by the closed nature of the nutrient cycle (a). Once the vegetation is removed (b), nutrients are quickly removed from the system creating infertile conditions, even deserts.

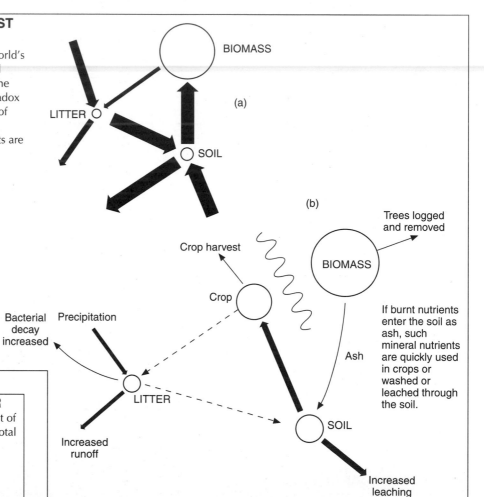

If burnt nutrients enter the soil as ash, such mineral nutrients are quickly used in crops or washed or leached through the soil.

ACRES LOST EACH YEAR

		Per cent of world total
Brazil	8.5 million	0.7%
Indonesia	3.0 million	1.0%
Bolivia	2.5 million	2.1%
Mexico	2.3 million	1.9%
Venezuela	2.1 million	1.5%
Zaire	1.8 million	0.6%

- Annual loss of rainforest is 40 million acres, the size of England and Wales:
    - Latin America   20.0 million acres
    - Africa   11.0 million acres
    - Asia   9.0 million acres
- Rainforest is disappearing at the rate of an acre per second.
- Rainforests support up to 90% of all wildlife.
- Annual deforestation in the 1990s was 50% higher than in the 1980s.
- Over 200 million people live in rainforest areas.

MAIN CONFLICTS

Cattle ranching	South America
Banana plantations	Costa Rica, South America
Coffee plantations	Africa
Logging	All over
Farming	All over
Mining	South America, Asia, Africa
Rubber plantations	Indonesia

Savannas

CLIMATE

- There is no such thing as a 'typical' savanna climate.
- Rainfall in savanna areas ranges from 500-2000 mm per year with a drought lasting between one and eight months.
- Annual temperatures are high (>25°C).
- Summers are hot and wet, winters hot and dry.
- Convectional rain occurs in summer.
- High temperatures year-round lead to high evapotranspiration losses.

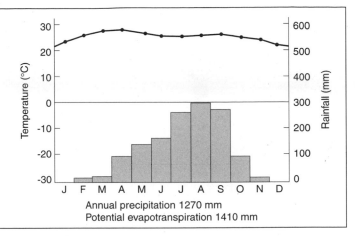

Annual precipitation 1270 mm
Potential evapotranspiration 1410 mm

SOILS: FERRUGINOUS

Savanna soils are influenced by distinct seasonal changes in processes. Moreover, they vary with topography. Frequently, sandy and/or leached soils predominate on the upper slopes, clay-based soils on lower slopes. This **catena** is reflected by changes in vegetation. In places **laterite**, a hardened layer of iron/aluminium, may limit further soil development and agricultural practices.

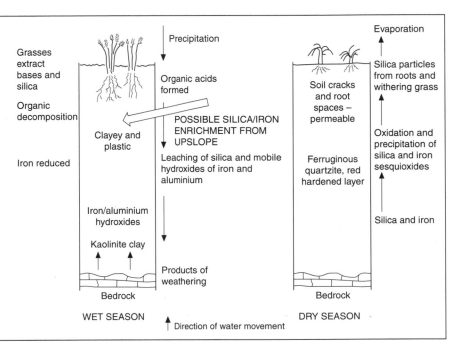

VEGETATION

Savanna vegetation is **xerophytic** (adapted to drought) and **pyrophytic** (adapted to fire). Grasses predominate on sandy, leached soils, while trees may be found in moister areas, such as valleys. Growth is rapid in the summer (NPP 900 g/m²/yr).

Fire (natural and as a result of human activity) reduces the biomass store (4 kg/m). Grasses are well adapted because the bulk of their biomass is beneath ground level and they regenerate quite quickly after burning.

NUTRIENT CYCLE

- The biomass store is less than that of the tropical rainforest due to a shorter growing season.
- The litter store is small due to fire. This means that the soil store is relatively large.

The savanna nutrient cycle differs from the tropical rainforest nutrient cycle due to the combined effects of the seasonal drought and the occurrence of fire. Consequently there is:
  (i) a lower nutrient availability
  (ii) a reduced biomass store
  (iii) a small litter store
  (iv) a relatively large soil store

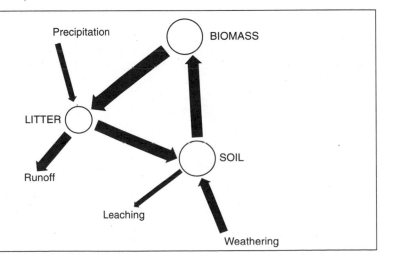

Temperate grasslands (steppe)

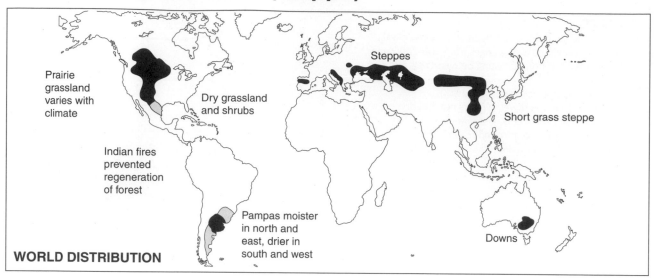

WORLD DISTRIBUTION

Prairie grassland varies with climate

Dry grassland and shrubs

Indian fires prevented regeneration of forest

Pampas moister in north and east, drier in south and west

Steppes

Short grass steppe

Downs

CLIMATE

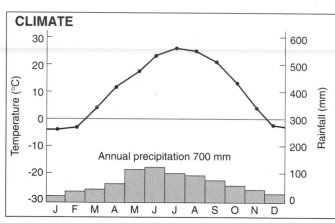

Annual precipitation 700 mm

- Climate is continental in character, i.e. hot summers, up to 30°C, and cold winters, with temperatures below freezing for up to 6 months a year.
- Precipitation is low (250-750 mm per year), with snow in winter and convectional rain in summer.
- Evaporation rates are high, especially in summer.
- In the USA, precipitation decreases from east to west, e.g. Virginia and Tennessee 750 mm, Nevada 250 mm.

SOIL: CHERNOZEM

Grasslands are dominated by black earth or **chernozem** soils. These are frequently developed on wind-blown loess deposits, rich in **calcium carbonate** ($CaCO_3$). Burrowing animals and soil fauna mix the soil. There may be some concentration of $CaCO_3$ due to spring leaching (snowmelt) and summer capillary rise. In the USA, soils vary from east to west. In the eastern areas of higher rainfall, brown earths are formed, in the drier regions of the west, chestnut soils are found.

A

Dark colour – much organic matter

Nodules of $CaCO_3$

1.5 m

C

Slight leaching due to snowmelt and thunderstorms; capillary rise in summer

NUTRIENT CYCLE

The natural nutrient cycle is affected by:
- climate (drought)
- fire
- the highly matted root system
- the grassland vegetation

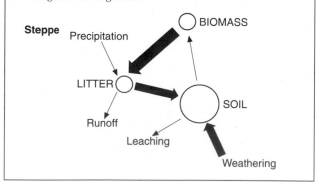

Steppe

Precipitation

BIOMASS

LITTER

Runoff

Leaching

SOIL

Weathering

VEGETATION

Vegetation is closely related to climate:
- trees in eastern (wetter) areas and along water courses
- height of grass relates to amount of precipitation

Tall grass, e.g. Bluestem (1.5-2.5 m), is found in the east where precipitation is about 750 mm per year.

Mixed grass, e.g. Little Bluestem (0.6-1.5 m), is found in central areas where precipitation is about 400-600 mm per year.

Short bunch, e.g. Buffalo Grass (<0.5 m), is found in the west where precipitation is about 250 mm per year.

Vegetation is both **xerophytic** and **pyrophytic**.
- NPP is 600 g/m^2/yr
- Biomass 1.6 kg/m (due to the lack of woody species)

Temperate deciduous woodland

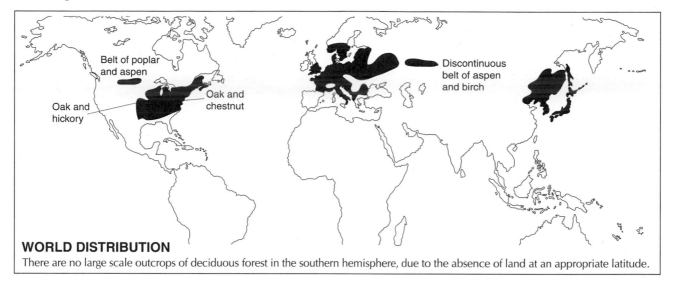

WORLD DISTRIBUTION

There are no large scale outcrops of deciduous forest in the southern hemisphere, due to the absence of land at an appropriate latitude.

CLIMATE

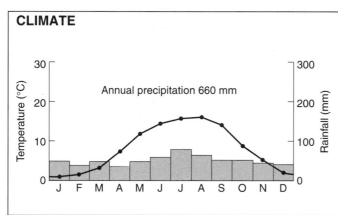

- There are wide variations in climate, e.g. between North-East USA and South-West Ireland.
- Rainfall is 500-1500 mm per year, with a winter maximum. It is mostly frontal (cyclonic) rainfall.
- Precipitation is greater than evapotranspiration, e.g. in Ireland, precipitation is 1000 mm per year and evapotranspiration is 450 mm per year.
- Winters below freezing (for 2-3 months in eastern China and North-East USA), although they are milder in western Europe due to the Gulf Stream.
- Cool summers, 15-20°C.

SOIL: BROWN EARTH

- Soils are generally quite fertile.
- The mull humus is mildly acidic (pH 5.5-6.5).
- Soil fauna such as earthworms flourish, mixing the layers and nutrients.
- Decomposition of leaves takes up to 9 months.
- Blurred horizons due to earthworm activity.

O	L F H	Mull humus, mildly acidic
A	Brown	
E	Light colour (leached of minerals)	Slight leaching of Ca, Mg, Fe, and Al
B	Deposition of clay, humus, and iron	Illuviation
C		Bedrock

NUTRIENT CYCLE

There is a large store of nutrients in the soil due to:
- slow growth in winter
- a low density of vegetation compared with the tropical rainforest

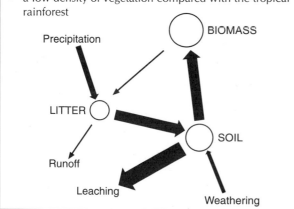

VEGETATION

- **Deciduous** trees shed their leaves in winter to retain moisture, conserve nutrients, and avoid damage by snow/ice.
- **NPP** is high - 1200 g/m^2/year due to high summer temperatures and the long hours of daylight.
- **Biomass** is high, 35 kg/m^2, due to large amounts of woody material.
- Vegetation varies with **soil** type, e.g. on more acidic soils birch and rowan are found, on more alkaline soils box and maple (oak is a generalist).
- Shrub vegetation varies with light, e.g. **heliophytes** need light, such as the wood anemone which flowers early, while **sciophytes** tolerate darkness, e.g. dog's mercury and ivy.

Temperate coniferous (boreal) forest

WORLD DISTRIBUTION

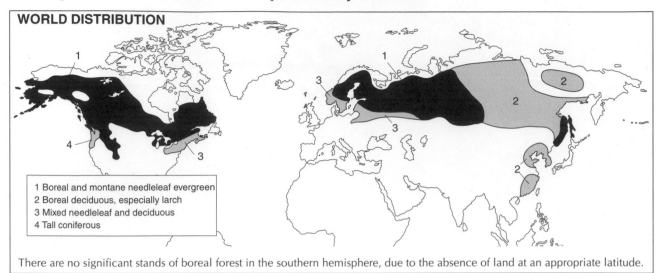

1 Boreal and montane needleleaf evergreen
2 Boreal deciduous, especially larch
3 Mixed needleleaf and deciduous
4 Tall coniferous

There are no significant stands of boreal forest in the southern hemisphere, due to the absence of land at an appropriate latitude.

CLIMATE

Annual precipitation 320 mm

J F M A M J J A S O N D

The climate is a cool temperate or cold continental climate, depending on the proximity of the ocean.

- Rainfall is low (<500 mm per year), with no real seasonal pattern. Snowfall is frequent in winter, and frosts occur in summer.
- Summer temperatures 10-15°C, with winter temperatures below freezing for up to 6 months.
- Growing season is limited to a maximum of 6-8 months, but the long hours of daylight in summer (16-20 hours) allow photosynthesis.

SOIL: PODZOL

- Precipitation greater than potential evapotranspiration.
- Leaching by snowmelt.
- Raw acid mor humus (pH 4.5-5.5).
- Acid water releases iron and aluminium oxides, transfers them down the soil, and may form an impermeable iron pan.
- Very few earthworms, due to acidity, and therefore little mixing of horizons.
- Thick litter layer due to the low temperatures, and resistant acidic nature of needles.

Thin organic layer — Organic matter — Silica

Ash-grey alleviated horizon — Ea

Dark brown depositer layer containing humus, clay, Fe, and Al — Bfe

Bedrock — C

NUTRIENT CYCLE

Low temperatures and slow rates of weathering produce large stores of nutrients in the litter layer.

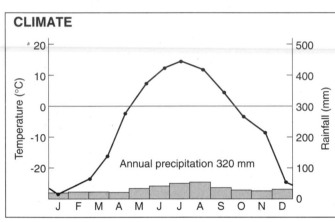

Precipitation

BIOMASS

LITTER

SOIL

Runoff

Leaching

Weathering

VEGETATION

- **Evergreen:** green throughout the year, and therefore able to potosynthesise when temperatures rise above 3°C.
- Conicle **shape** enables trees to shed snow and reduce rocking by wind.
- **Needle leaves** - small surface area, and therefore water loss is reduced.
- Generally occur in **stands** of one species.
- Pine favours sandy **soils**, spruce damper soils.
- Ground vegetation limited - it is too dark.
- NPP 800 g/m^2/yr.
- Biomass 20 kg/m (much woody matter).

Population growth

- This accelerating growth rate is a relatively recent phenomena.

- Growth rates doubled 1650-1850, 1850-1920, and 1920-1970.

- It is estimated that population will stabilise at around 12 billion; 95% of growth will be in less developed countries (LDCs).

- Large numbers of children have already been born in LDCs, so even if birth rate falls population will still grow (*population momentum*).

- The biggest relative increase will be in Africa (annual rise of 3%).

- By the end of the century Africa's 650 million will rise to 900 million - the population of Nigeria will double in the next 20 years.

- The result of the world's uneven population growth will be:
  (i) severe pressures on government stability (as well as services);
  (ii) increased tensions between 'haves' and 'have-nots';
  (iii) serious effects on food supply and the environment.

The rise of world population

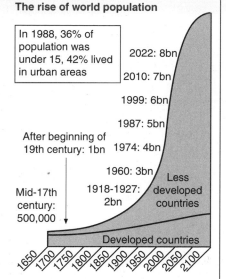

In 1988, 36% of population was under 15, 42% lived in urban areas

2022: 8bn
2010: 7bn
1999: 6bn
1987: 5bn
After beginning of 19th century: 1bn | 1974: 4bn
1960: 3bn
Mid-17th century: 500,000 | 1918-1927: 2bn
Less developed countries
Developed countries

1650 1700 1750 1800 1850 1900 1950 2000 2050 2100

PERIODS OF GROWTH (COWGILL, 1949)

CYCLE 1 **Malthusian**
- pre-industrial populations
- demographic variables and environmental conditions tend to check growth

Vital rates — BR
- - - DR
Temporary improvement

CYCLE 2 **Neolithic**
- agricultural revolution and reduced duration of breast feeding increases birth rate
- concentration in settlements leads to spread of diseases and increase in death rate

BR
DR

CYCLE 3 **Demographic transition**
- associated with industrialisation
- modern humankind is able to foresee demographic catastrophe before it arrives

DR
BR

CYCLE 4 **Baby boom**
- post-war period (1945-65)
- trend towards younger marriages and economic prosperity

Short-term increase
BR
DR

Japan has dropped below *replacement level* fertility. The result is an ageing populating:
- high dependency ratio
- increased pressure on services
- old people become politically powerful

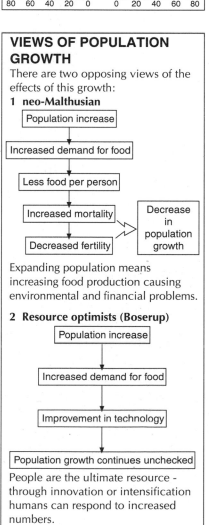

1980
- - 65 or over
Males   Females
Age category
85+
80–84
75–79
70–74
65–69
60–64
55–59
50–54
45–49
40–44
35–39
30–34
25–29
20–24
15–19
10–14
5–9
0–4
Persons (10,000)
80 60 40 20 0   0 20 40 60 80

2000 (estimate)
Males   Females
Age category
85+
80–84
75–79
70–74
65–69
60–64
55–59
50–54
45–49
40–44
35–39
30–34
25–29
20–24
15–19
10–14
5–9
0–4
Persons (10,000)
80 60 40 20 0   0 20 40 60 80

VIEWS OF POPULATION GROWTH

There are two opposing views of the effects of this growth:

1 neo-Malthusian

Population increase
↓
Increased demand for food
↓
Less food per person
↓
Increased mortality → Decrease in population growth
↓
Decreased fertility →

Expanding population means increasing food production causing environmental and financial problems.

2 Resource optimists (Boserup)

Population increase
↓
Increased demand for food
↓
Improvement in technology
↓
Population growth continues unchecked

People are the ultimate resource - through innovation or intensification humans can respond to increased numbers.

Population distribution and change

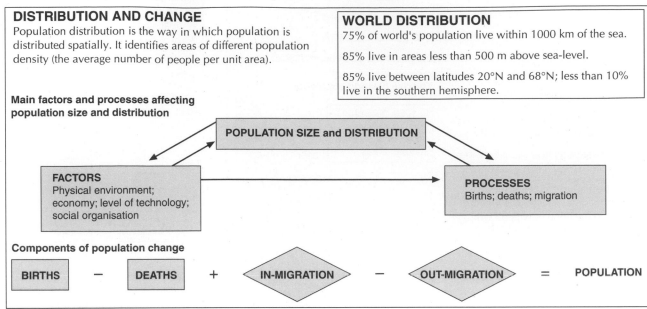

DISTRIBUTION AND CHANGE
Population distribution is the way in which population is distributed spatially. It identifies areas of different population density (the average number of people per unit area).

WORLD DISTRIBUTION
75% of world's population live within 1000 km of the sea.

85% live in areas less than 500 m above sea-level.

85% live between latitudes 20°N and 68°N; less than 10% live in the southern hemisphere.

Main factors and processes affecting population size and distribution

POPULATION SIZE and DISTRIBUTION

FACTORS
Physical environment; economy; level of technology; social organisation

PROCESSES
Births; deaths; migration

Components of population change

BIRTHS − DEATHS + IN-MIGRATION − OUT-MIGRATION = POPULATION

CASE STUDY: NATIONAL DISTRIBUTION IN THE UK, 1991

National level
- The South-East has one-third of the population living on a tenth of the land area.
- Scotland has one-tenth of the population living on one-third of the land area.
- The most congested region is the North-West with an average of 868 persons per square kilometre.

Low density
- Remote upland regions.
- Harsh physical conditions and a lack of economic opportunity.
- In parts of the Scottish Highlands there are large areas with no population.

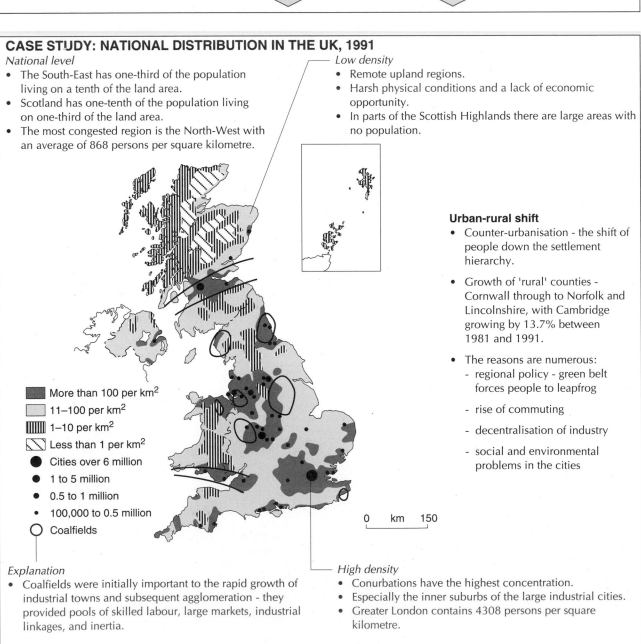

Key
- More than 100 per km²
- 11–100 per km²
- 1–10 per km²
- Less than 1 per km²
- ● Cities over 6 million
- ● 1 to 5 million
- ● 0.5 to 1 million
- · 100,000 to 0.5 million
- ○ Coalfields

0   km   150

Urban-rural shift
- Counter-urbanisation - the shift of people down the settlement hierarchy.
- Growth of 'rural' counties - Cornwall through to Norfolk and Lincolnshire, with Cambridge growing by 13.7% between 1981 and 1991.
- The reasons are numerous:
  - regional policy - green belt forces people to leapfrog
  - rise of commuting
  - decentralisation of industry
  - social and environmental problems in the cities

Explanation
- Coalfields were initially important to the rapid growth of industrial towns and subsequent agglomeration - they provided pools of skilled labour, large markets, industrial linkages, and inertia.

High density
- Conurbations have the highest concentration.
- Especially the inner suburbs of the large industrial cities.
- Greater London contains 4308 persons per square kilometre.

Fertility and mortality

FERTILITY

Fertility is the measured capacity of a population to generate births. Two measures are used:

- **crude birth rate** - the number of births in a given year divided by the population and multiplied by 1000 (reliant on accurate data and does not account for age-sex structure)

- **total fertility rate** - the number of children born to 1000 women passing through the child-bearing ages (assuming none of the women die); a fertility rate of 2.1 births per woman indicates replacement level (stable population)

Government policy	
Direct policies	**Indirect policies**
Policy and laws	Government spending
1. Minimum marriage age	1. Education
2. Women's status	2. Primary healthcare
3. Children's education and work	3. Family planning
4. Breast-feeding	4. Incentives for fertility control
5. Number of children per family	5. Old-age security
	Tax programmes
	1. Family allowances
	2. User fees for larger families

The value of children

	Benefits	Costs
Economic	Help with domestic chores	Cost of education
	Financial contribution to household	Cost of food, clothing, and shelter
	Security in old age	Loss of parental wage earnings
Social	Companionship, love, happiness	Mental strains
	Marital bonds strengthened	Overcrowding of family residence
	Continuation of family name	
Psychological	Fulfilment	Parents feeling tied down
	Living through children	Emotional strain
	Incentive to succeed	Disciplinary problems

Exposure to intercourse
- Age at marriage or sexual union
- Monogamy/polygamy
- Widowhood, divorce
- Spousal separation
- Coital frequency
- Post-birth abstinence

Conception
- Natural sterility
- Pathological sterility
- Lactational amenorrhea
- Contraceptive use

Pregnancy outcome
- Spontaneous abortion
- Induced abortion

Fertility (live births)

e.g. Government policies
Religious philosophy

Socio-economic determinants
e.g. Economic and social value of children
Economic and social status of women

Proximate determinants
e.g. Marriage patterns | Induced abortion
Patterns of sexual activity | Sterility
Length of breast-feeding | Usage of contraception

FERTILITY

What changes birth rates?	
Availability of family planning information and services	Migration to towns and cities
Education and literacy	A better deal for women
Better health and fewer child deaths	More employment opportunities
Later marriages	More equal income distribution and rising living standards

MORTALITY

Mortality is different from fertility as it tends to be more stable and predictable. There are a number of measures:

- **crude death rate** - the number of deaths in a specific period per 1000 of population (distorted by age structure)

- **age-specific mortality rates** - the number of deaths of people of a certain age per 1000, e.g. infant mortality rate

- **life expectancy at birth** - the average number of years a person can expect to live

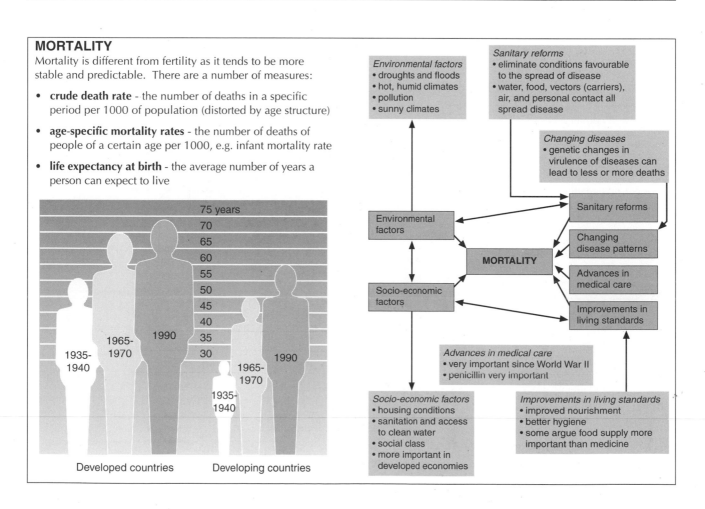

Environmental factors
- droughts and floods
- hot, humid climates
- pollution
- sunny climates

Sanitary reforms
- eliminate conditions favourable to the spread of disease
- water, food, vectors (carriers), air, and personal contact all spread disease

Changing diseases
- genetic changes in virulence of diseases can lead to less or more deaths

Environmental factors

Socio-economic factors

MORTALITY

Sanitary reforms

Changing disease patterns

Advances in medical care

Improvements in living standards

Advances in medical care
- very important since World War II
- penicillin very important

Socio-economic factors
- housing conditions
- sanitation and access to clean water
- social class
- more important in developed economies

Improvements in living standards
- improved nourishment
- better hygiene
- some argue food supply more important than medicine

Developed countries   Developing countries

The demographic transition model

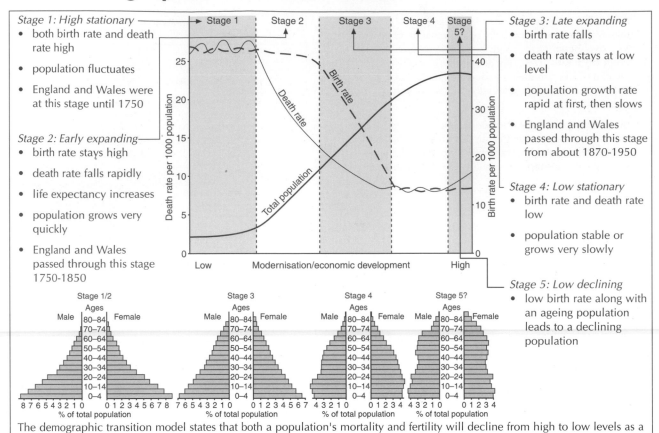

Stage 1: High stationary
- both birth rate and death rate high
- population fluctuates
- England and Wales were at this stage until 1750

Stage 2: Early expanding
- birth rate stays high
- death rate falls rapidly
- life expectancy increases
- population grows very quickly
- England and Wales passed through this stage 1750-1850

Stage 3: Late expanding
- birth rate falls
- death rate stays at low level
- population growth rate rapid at first, then slows
- England and Wales passed through this stage from about 1870-1950

Stage 4: Low stationary
- birth rate and death rate low
- population stable or grows very slowly

Stage 5: Low declining
- low birth rate along with an ageing population leads to a declining population

The demographic transition model states that both a population's mortality and fertility will decline from high to low levels as a result of social and economic development. It is based on the European experience, but has also been used a a predictive tool to explain change in the developing world.

DEMOGRAPHIC TRANSITION AND GLOBAL TRENDS

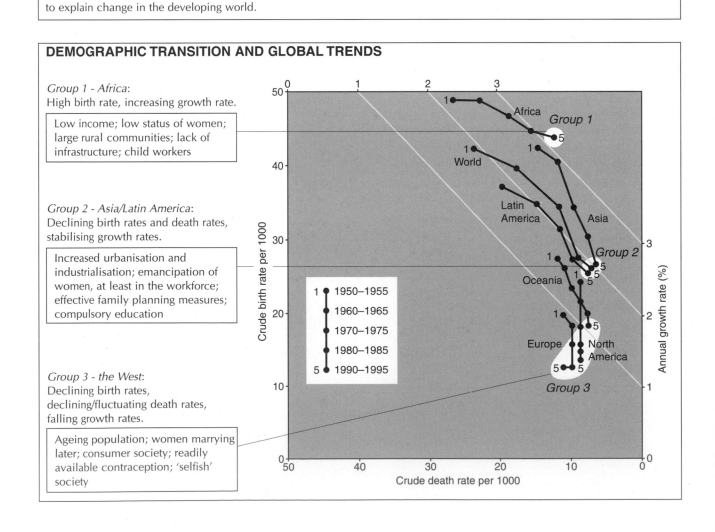

Group 1 - Africa:
High birth rate, increasing growth rate.

Low income; low status of women; large rural communities; lack of infrastructure; child workers

Group 2 - Asia/Latin America:
Declining birth rates and death rates, stabilising growth rates.

Increased urbanisation and industrialisation; emancipation of women, at least in the workforce; effective family planning measures; compulsory education

Group 3 - the West:
Declining birth rates, declining/fluctuating death rates, falling growth rates.

Ageing population; women marrying later; consumer society; readily available contraception; 'selfish' society

Optimum population and population policy

OPTIMUM POPULATION

Optimum population
- the size of population which permits the full utilisation of the natural resources of an area giving maximum per capita output and standard of living

Underpopulation
- population is too small to develop its resources effectively

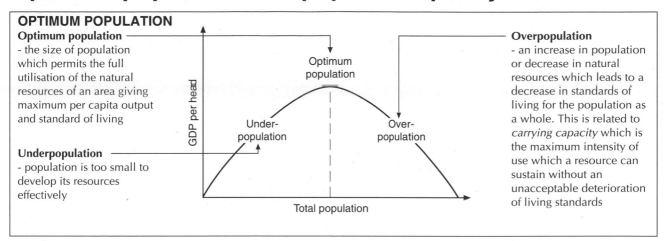

Overpopulation
- an increase in population or decrease in natural resources which leads to a decrease in standards of living for the population as a whole. This is related to *carrying capacity* which is the maximum intensity of use which a resource can sustain without an unacceptable deterioration of living standards

POPULATION-RESOURCE REGIONS (ACKERMAN)

Optimum population is based on the combination of three factors: **population density, resources,** and **technology**. Regions of the world can be classified according to population-resource ratios.

Ackerman's classification shows many of the problems associated with the concept of optimum population:

- only measured in economic terms

- little time given to social or regional inequalities within a country

- population structure (active population, dependency ratio) is not considered

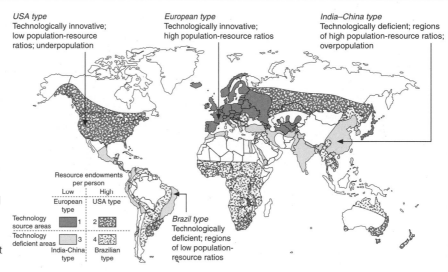

CASE STUDY: POPULATION POLICY IN CHINA

China has used the concept of optimum population to stabilise its population at 1.2 billion by the year 2000 and to reduce the population to a government-set optimum of 700 million within a century. A number of different policies have been initiated to achieve this.

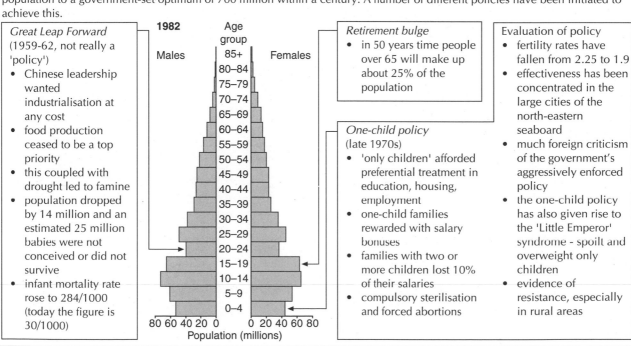

Great Leap Forward (1959-62, not really a 'policy')
- Chinese leadership wanted industrialisation at any cost
- food production ceased to be a top priority
- this coupled with drought led to famine
- population dropped by 14 million and an estimated 25 million babies were not conceived or did not survive
- infant mortality rate rose to 284/1000 (today the figure is 30/1000)

Retirement bulge
- in 50 years time people over 65 will make up about 25% of the population

One-child policy (late 1970s)
- 'only children' afforded preferential treatment in education, housing, employment
- one-child families rewarded with salary bonuses
- families with two or more children lost 10% of their salaries
- compulsory sterilisation and forced abortions

Evaluation of policy
- fertility rates have fallen from 2.25 to 1.9
- effectiveness has been concentrated in the large cities of the north-eastern seaboard
- much foreign criticism of the government's aggressively enforced policy
- the one-child policy has also given rise to the 'Little Emperor' syndrome - spoilt and overweight only children
- evidence of resistance, especially in rural areas

Migration

DEFINITION

There are various forms of spatial mobility, but not all are considered migration. Population movements take four forms:

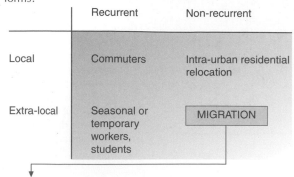

	Recurrent	Non-recurrent
Local	Commuters	Intra-urban residential relocation
Extra-local	Seasonal or temporary workers, students	MIGRATION

Migration is defined as a movement involving a change in permanent residence with a complete readjustment of the community affiliations of the migrant.

CLASSIFICATIONS

Migration can be classified in terms of distance (internal or international), time (temporary or permanent), and origin (rural or urban).

Conditions which cause migration can involve both 'push' and 'pull' factors.

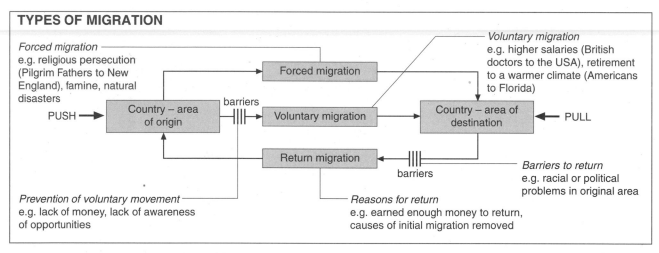

Push: Poor employment, Low income, Housing shortages, Intolerance, Adverse climatic conditions, Natural disasters, Social upheaval

Pull: Job prospects, High wages, Improved housing, Tolerance, Amenities, Attractive environment, High standard of living

TYPES OF MIGRATION

Forced migration
e.g. religious persecution (Pilgrim Fathers to New England), famine, natural disasters

PUSH → Country – area of origin — barriers → Forced migration / Voluntary migration / Return migration → Country – area of destination ← PULL

Voluntary migration
e.g. higher salaries (British doctors to the USA), retirement to a warmer climate (Americans to Florida)

Barriers to return
e.g. racial or political problems in original area

Prevention of voluntary movement
e.g. lack of money, lack of awareness of opportunities

Reasons for return
e.g. earned enough money to return, causes of initial migration removed

THE IMPACT OF MIGRATION

(a) Rural-urban migration in Zambia

Benefits	-to the destination	-for the migrants
	• young male labour	• labour rates higher
	• centralisation of population	• amenities

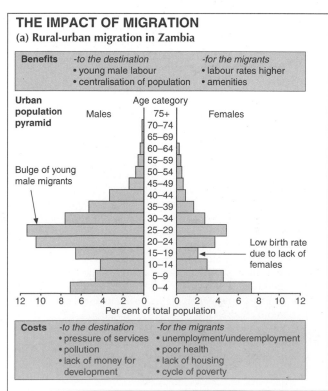

Urban population pyramid

Age category (Males / Females): 75+, 70–74, 65–69, 60–64, 55–59, 50–54, 45–49, 40–44, 35–39, 30–34, 25–29, 20–24, 15–19, 10–14, 5–9, 0–4

Bulge of young male migrants

Low birth rate due to lack of females

Per cent of total population (12 10 8 6 4 2 0 — 0 2 4 6 8 10 12)

Costs	-to the destination	-for the migrants
	• pressure of services	• unemployment/underemployment
	• pollution	• poor health
	• lack of money for development	• lack of housing
		• cycle of poverty

(b) Turkish immigrants in the Netherlands

Benefits	-to the destination	-for the migrants
	• cheap labour willing to do low-paid jobs	• chance of Western-style life
		• better education and health care

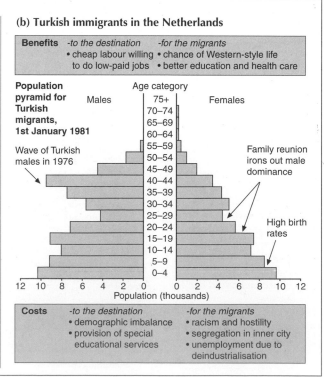

Population pyramid for Turkish migrants, 1st January 1981

Age category (Males / Females): 75+, 70–74, 65–69, 60–64, 55–59, 50–54, 45–49, 40–44, 35–39, 30–34, 25–29, 20–24, 15–19, 10–14, 5–9, 0–4

Wave of Turkish males in 1976

Family reunion irons out male dominance

High birth rates

Population (thousands) (12 10 8 6 4 2 0 — 0 2 4 6 8 10 12)

Costs	-to the destination	-for the migrants
	• demographic imbalance	• racism and hostility
	• provision of special educational services	• segregation in inner city
		• unemployment due to deindustrialisation

Site and situation

SITE

The actual location on which dwellings are erected. The area occupied by a settlement. Site characteristics include:

- well-drained, free from flooding (dry point)
- reliable water supply (wet point)
- defence, water, fuel, building materials, land for crops and livestock

SITUATION

The location of a settlement relative to other settlements and the surrounding physical geography (links to routeways and trade). Situation characteristics include:

- the convergence of valleys
- a gap in a ridge or range of hills
- a bridging point (especially lowest bridging point)

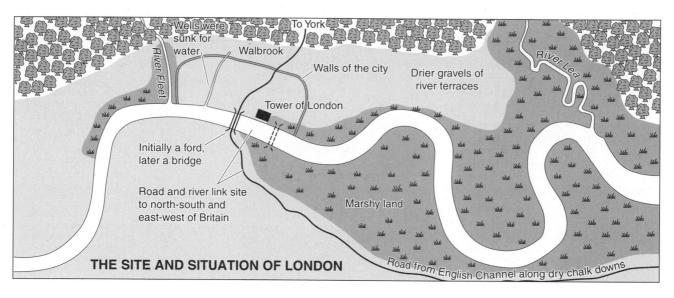

THE SITE AND SITUATION OF LONDON

Wells were sunk for water

To York

Walbrook

River Fleet

Walls of the city

Drier gravels of river terraces

River Lea

Tower of London

Initially a ford, later a bridge

Road and river link site to north-south and east-west of Britain

Marshy land

Road from English Channel along dry chalk downs

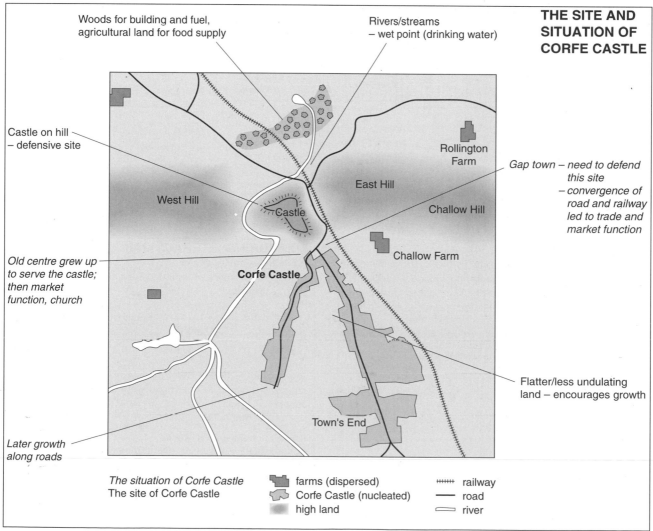

THE SITE AND SITUATION OF CORFE CASTLE

Woods for building and fuel, agricultural land for food supply

Rivers/streams – wet point (drinking water)

Castle on hill – defensive site

Rollington Farm

Gap town – need to defend this site – convergence of road and railway led to trade and market function

West Hill

East Hill

Challow Hill

Castle

Old centre grew up to serve the castle; then market function, church

Challow Farm

Corfe Castle

Flatter/less undulating land – encourages growth

Later growth along roads

Town's End

The situation of Corfe Castle
The site of Corfe Castle

- farms (dispersed)
- Corfe Castle (nucleated)
- high land
- railway
- road
- river

Central place theory (CPT)

CHRISTALLER'S MODEL (1933)

A model is a simplification of reality. It identifies a number of theories or ideas which may affect the geographical landscape. It is important to grasp that it is the *conceptual framework* rather than the *initial assumptions* (which simplify the real world) which is important. There are a number of concepts which Christaller identified in determining the size and spacing of settlements.

Threshold and range

Threshold population is the minimum sales that an establishment must secure in order to survive.

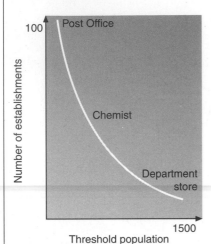

Range is the maximum distance the dispersed population is prepared to travel to purchase goods from a central place.

Range must always be greater than the area enclosing the threshold population.

Comparison goods are high order, high cost, and not frequently purchased. They have a high threshold and a large range.

Convenience goods are low order, low cost, and frequently purchased. They have a low threshold and short range.

Hierarchy

Hierarchy has three elements:

1 Higher order settlements offer more functions and are normally larger in population size.

2 Higher order settlements occur less frequently, are spaced more widely apart, and serve larger trade areas.

3 The hierarchy is 'stepped'.

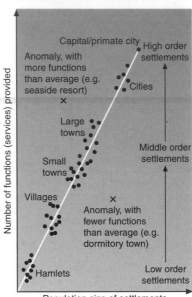

At each level the service or market area of each settlement will be different. Influences include price of a good, number/density of inhabitants, competition from other settlements, and income/social status.

Nesting

Each settlement is surrounded by a market area which it serves. Larger settlements have larger market areas. In CPT market areas are hexagonal and their nesting determines their size and orientation.

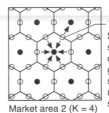

Market optimising Shoppers in the smaller settlements divide into 3 equal groups when shopping in the 3 nearest larger settlements

Market area 1 (K = 3)

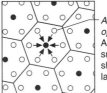

Transport optimising Shoppers in the smaller settlements divide into 2 equal groups when shopping in the 2 nearest larger settlements

Market area 2 (K = 4)

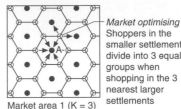

Administration optimising All shoppers in the smaller settlements shop in the nearest large settlement

Market area 3 (K = 7)

K=3 gives the greatest choice of settlement to each customer.

K=7 gives the highest degree of centralised control to the central place.

THE LÖSCH MODEL (1954)

Lösch modified Christaller's model by producing many more hexagons. He rotated the hexagons to produce a landscape very different to that of Christaller. He produced clusters of high order settlements rather than the original evenly spaced settlements.

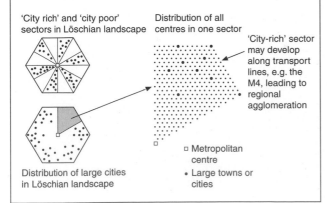

'City rich' and 'city poor' sectors in Löschian landscape

Distribution of all centres in one sector

'City-rich' sector may develop along transport lines, e.g. the M4, leading to regional agglomeration

Distribution of large cities in Löschian landscape

□ Metropolitan centre

• Large towns or cities

MEASURING INTERACTION

(a) Reilly's Gravity Model

Calculates the sphere of influence of a central place:

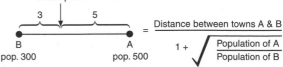

$$\text{Break point} = \frac{\text{Distance between towns A \& B}}{1 + \sqrt{\dfrac{\text{Population of A}}{\text{Population of B}}}}$$

'Break point' is determined by population and distance. However, population does not always indicate number of services, e.g. tourist towns have a small resident population but a large number of services.

(b) Huff's Behavioural Model

Calculates the probability of a customer buying goods in his or her town rather than another (P_1), i.e.

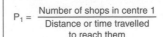

$$P_1 = \frac{\text{Number of shops in centre 1}}{\text{Distance or time travelled to reach them}} \div \frac{\text{Total number of shops in the whole study area}}{\text{Distance or time travelled to reach them}}$$

Rural land use

DEFINITION OF RURAL
Less densely populated parts of a country which are recognised by their visual 'countryside' components. Three criteria are used:
- economic - a high dependence on agriculture for income
- social and demographic - the 'rural way of life' and low population density
- geographical - remoteness from urban centres

TYPES OF RURAL AREA
- Extensive land use where there is little demand for land, e.g. Highlands and Islands of Scotland.
- Intensive land use in the developing world as a result of overpopulation.
- In the West, high land values and negative externalities in urban areas has led to the 'suburbanisation' of rural areas.

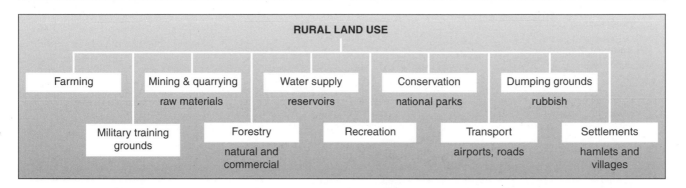

RURAL LAND USE

| Farming | Mining & quarrying | Water supply | Conservation | Dumping grounds |
| | raw materials | reservoirs | national parks | rubbish |

| Military training grounds | Forestry | Recreation | Transport | Settlements |
| | natural and commercial | | airports, roads | hamlets and villages |

THE RURAL-URBAN FRINGE
Characteristics of changing communities in the rural-urban fringe include:
- **segregation** of rural areas into large blocks of one class or price of housing
- **selective immigration** of mobile middle-classes who live and work in distinct and separate social and economic worlds
- **commuting** of middle class workers

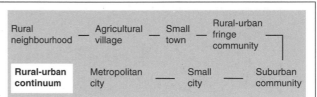

| Rural neighbourhood | — | Agricultural village | — | Small town | — | Rural-urban fringe community |
| **Rural-urban continuum** | | Metropolitan city | — | Small city | — | Suburban community |

- **collapse of geographical and social hierarchies** as the community becomes more outward-looking and the 'squirarchy' is replaced by a class-based structure polarised around housing segregation

WAUGH'S MODEL OF RURAL SETTLEMENTS

Linear development
The tendency for large towns and cities to grow outwards along roads has been checked by the **green belt** policy.

New or overspill towns
Designed both to accommodate the urban workforce and to act as self-supporting alternatives.

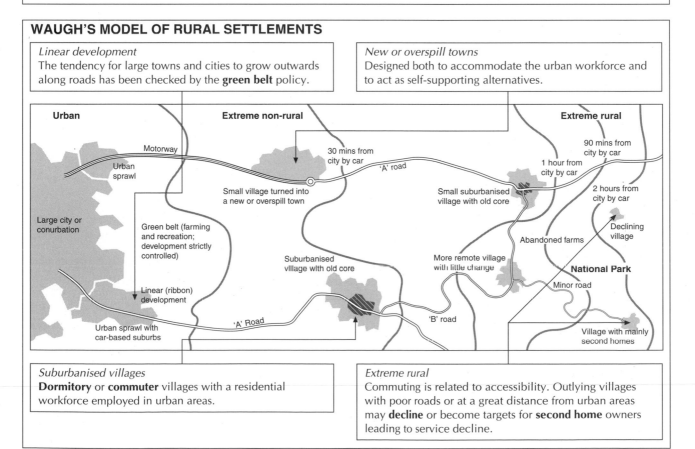

Suburbanised villages
Dormitory or **commuter** villages with a residential workforce employed in urban areas.

Extreme rural
Commuting is related to accessibility. Outlying villages with poor roads or at a great distance from urban areas may **decline** or become targets for **second home** owners leading to service decline.

Rural change

THE LEWIS AND MAUNDER MODEL (1970)

The model describes the changes in rural population over the last 200 years. The changes are related to three time periods. Each time period contains one or more 'landscapes' which relate to processes of migration and socio-economic change.

Landscapes

I   *Traditional communities*

II *Depopulation* (absolute population decrease) - due to:

  - reduction in farm employment
  - rural-urban migration
  - deterioration in age structure
  - reduced natural increase

III *Repopulation* (urban-rural migration) - due to:

  - decentralisation of industry
  - increased mobility
  - retirement
  - negative externalities of urban areas

IV *Population* (metropolitan expansion) - due to:

  - gentrification
  - urban policy

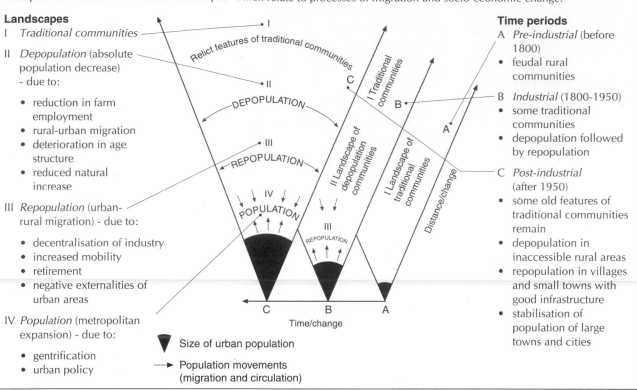

Size of urban population

→ Population movements (migration and circulation)

Time periods

A *Pre-industrial* (before 1800)
  - feudal rural communities

B *Industrial* (1800-1950)
  - some traditional communities
  - depopulation followed by repopulation

C *Post-industrial* (after 1950)
  - some old features of traditional communities remain
  - depopulation in inaccessible rural areas
  - repopulation in villages and small towns with good infrastructure
  - stabilisation of population of large towns and cities

THE ROBERTS MODEL (1989)

This model attempts to explain village and hamlet development in Western Europe. It identifies *the visible* (the landscape lived in and used) and *rural processes* (at work causing change).

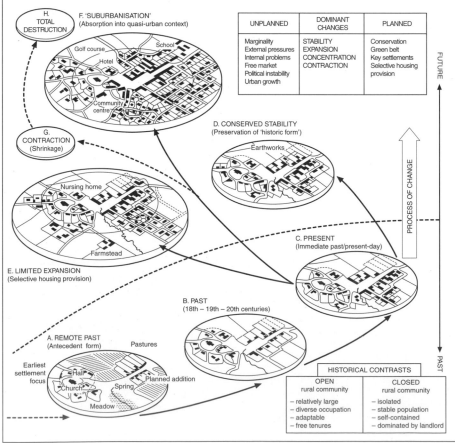

	UNPLANNED	DOMINANT CHANGES	PLANNED
	Marginality	STABILITY	Conservation
	External pressures	EXPANSION	Green belt
	Internal problems	CONCENTRATION	Key settlements
	Free market	CONTRACTION	Selective housing provision
	Political instability		
	Urban growth		

HISTORICAL CONTRASTS	
OPEN rural community	CLOSED rural community
– relatively large	– isolated
– diverse occupation	– stable population
– adaptable	– self-contained
– free tenures	– dominated by landlord

In the model, A, B, and C are similar to the three time periods in the Lewis and Maunder model; D, E, F, and G explain possible futures.

A **Remote past** - process of *concentration* related to the appearance of planned growth.

B **Past** - processes of *expansion*, *contraction*, or *stability* depending on socio-economic factors.

C **Present** - *closed* or *open* communities; changes due to *planning decisions* or from forces deriving from *modern society*.

D **Conserved stability** - enforced stability due to historic conservation.

E **Limited expansion** - growth 'in character', i.e. using materials and styles which blend with existing buildings.

F **Suburbanisation** - commuters give rise to modern houses, shops, and services providing for an urban community.

G **Contraction** - as certain villages are chosen as growth areas, other less accessible areas may decline or be abandoned as agriculture declines or is mechanised.

CASE STUDY: SMALL VILLAGES ON THE GOWER

The Gower Peninsula is in south-west Wales, to the west of Swansea. It is an Area of Outstanding Natural Beauty forming the north-westerly limit of the Bristol Channel. In the past thirty years the rural area of Gower has seen dramatic changes in the social and economic characteristics of its population. Processes of repopulation and depopulation are apparent in the villages of the Gower. Growth and decline are based on closeness to Swansea, accessibility, planning decisions, and natural beauty.

Small villages on the Gower Peninsula

The nine small villages

Overspill villages (suburbanised expansion due to out-migration from Swansea)

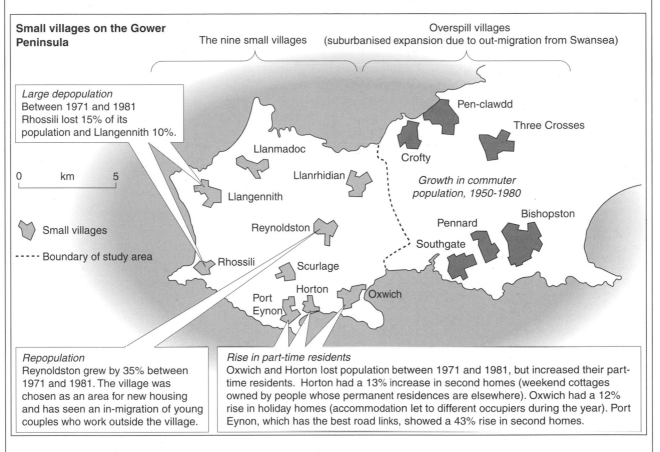

Large depopulation
Between 1971 and 1981 Rhossili lost 15% of its population and Llangennith 10%.

0   km   5

◆ Small villages

- - - - Boundary of study area

Growth in commuter population, 1950-1980

Repopulation
Reynoldston grew by 35% between 1971 and 1981. The village was chosen as an area for new housing and has seen an in-migration of young couples who work outside the village.

Rise in part-time residents
Oxwich and Horton lost population between 1971 and 1981, but increased their part-time residents. Horton had a 13% increase in second homes (weekend cottages owned by people whose permanent residences are elsewhere). Oxwich had a 12% rise in holiday homes (accommodation let to different occupiers during the year). Port Eynon, which has the best road links, showed a 43% rise in second homes.

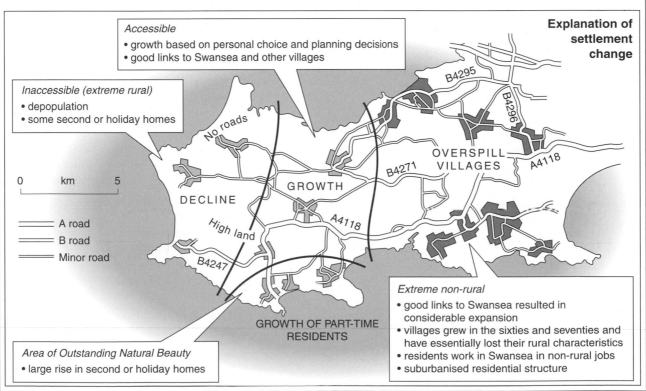

Explanation of settlement change

Accessible
- growth based on personal choice and planning decisions
- good links to Swansea and other villages

Inaccessible (extreme rural)
- depopulation
- some second or holiday homes

0   km   5

═══ A road
═══ B road
═══ Minor road

Area of Outstanding Natural Beauty
- large rise in second or holiday homes

Extreme non-rural
- good links to Swansea resulted in considerable expansion
- villages grew in the sixties and seventies and have essentially lost their rural characteristics
- residents work in Swansea in non-rural jobs
- suburbanised residential structure

Residential land use

There are a number of important models of urban structure.

CONCENTRIC ZONE MODEL (BURGESS, 1925)

Key for all diagrams

1. CBD
2. Zone in transition/light manufacturing
3. Low-class residential
4. Medium-class residential
5. High-class residential
6. Heavy manufacturing
7. Outlying business district
8. Residential suburb
9. Industrial suburb
10. Commuter zone

Building age decreases outwards

- model based on Chicago in the 1920s
- the city is growing spatially due to immigration and natural increase
- the area around the CBD has the lowest status and highest density housing
- residents move outwards with increasing social class and their homes are taken by new migrants

MULTIPLE-NUCLEI MODEL (HARRIS AND ULLMAN, 1945)

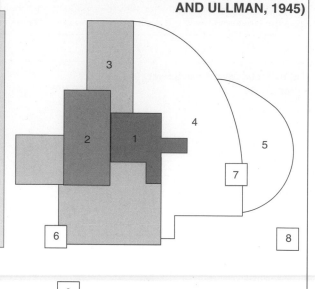

- model states that cities are not organised around one centre
- cities grow and embrace a number of centres
- other centres may develop with different commercial, administrative, and service centres

SECTOR MODEL (HOYT, 1939)

- sectors develop along routeways and growth is by extension of the sector
- high status residential areas are located away from industry

MURDIE'S (1969) IDEALISED MODEL OF URBAN ECOLOGICAL STRUCTURE

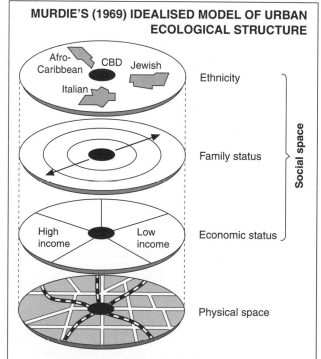

- identifies social, economic, and ethnic areas
- the 'residential mosaic' is superimposed on the physical space
- economic status has a sector pattern, family status has a concentric pattern, ethnic groups occur in clusters

FILTERING AND GENTRIFICATION

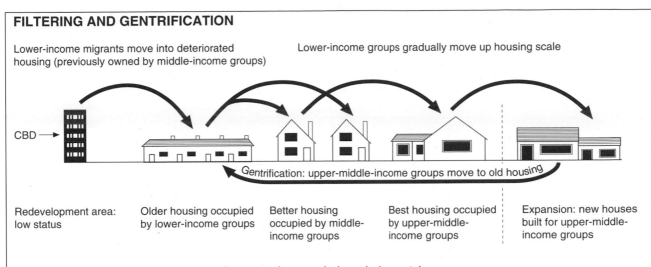

Lower-income migrants move into deteriorated housing (previously owned by middle-income groups)

Lower-income groups gradually move up housing scale

CBD →

Gentrification: upper-middle-income groups move to old housing

| Redevelopment area: low status | Older housing occupied by lower-income groups | Better housing occupied by middle-income groups | Best housing occupied by upper-middle-income groups | Expansion: new houses built for upper-middle-income groups |

Filtering occurs as housing deteriorates and it moves downwards through the social groups.
Gentrification reverses this process as middle-income groups upgrade older city properties by renovating them.

FAMILY LIFE-CYCLE MODEL

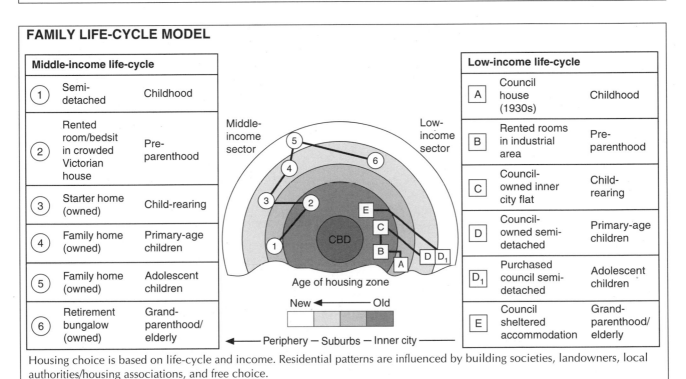

Middle-income life-cycle

1	Semi-detached	Childhood
2	Rented room/bedsit in crowded Victorian house	Pre-parenthood
3	Starter home (owned)	Child-rearing
4	Family home (owned)	Primary-age children
5	Family home (owned)	Adolescent children
6	Retirement bungalow (owned)	Grand-parenthood/elderly

Low-income life-cycle

A	Council house (1930s)	Childhood
B	Rented rooms in industrial area	Pre-parenthood
C	Council-owned inner city flat	Child-rearing
D	Council-owned semi-detached	Primary-age children
D_1	Purchased council semi-detached	Adolescent children
E	Council sheltered accommodation	Grand-parenthood/elderly

Middle-income sector · Low-income sector · CBD · Age of housing zone

New ← → Old

← Periphery — Suburbs — Inner city →

Housing choice is based on life-cycle and income. Residential patterns are influenced by building societies, landowners, local authorities/housing associations, and free choice.

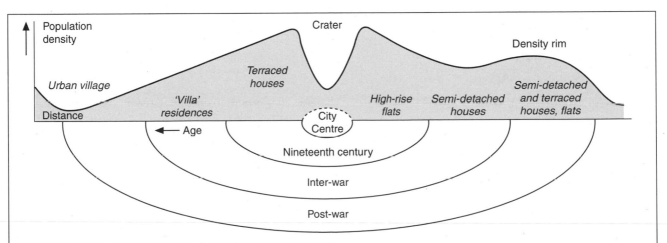

Population density

Crater

Density rim

Urban village

Terraced houses

'Villa' residences

High-rise flats

Semi-detached houses

Semi-detached and terraced houses, flats

Distance

← Age

City Centre

Nineteenth century

Inter-war

Post-war

THE DENSITY CURVE WITH A HIGH-DENSITY RIM

There are low residential densities in the city centre. Peak residential densities are beyond the central area, in the inner city. The rise in densities in the outer suburbs is due to the development of council housing estates in the last thirty years.

The central business district (CBD)

The central business district (CBD) is generally at the heart of the city and is the focus for the urban transport system. It is the centre of the city's commercial, social, and cultural life. It possesses a number of clearly defined characteristics.

CHARACTERISTICS OF THE CBD

Absence of manufacturing
- a few specialised activities such as newspapers

Concentration of offices
- central location needed for clients and workforce
- offices tend to locate in zones

Multi-storey development
- high land values encourage buildings to grow upwards
- the real area of the CBD should therefore be measured in floor space rather than ground space

Vertical zoning
- land use changes within multi-storey blocks
- shops and services occupy ground floors

Low residential population
- high bid rents mean that few people live in the CBD
- there may be some luxury flats, especially in large cities

Pedestrianisation
- since the 1960s urban traffic management has limited the movement of vehicles within the CBD
- pedestrianisation has made shopping safer but some town centres have lost character

Concentration of retailing
- accessibility attracts shops with wide ranges and high threshold populations
- department stores and high threshold chains dominate in the centre
- specialist shops occupy less accessible sites

Comprehensive redevelopment
- the clearance of sites for complete rebuilding
- sometimes the CBD is extended into the inner city, causing conflict with residents
- redevelopment can also shift the centre of the CBD, causing some businesses to decline

CASE STUDY: THE CBD OF SWANSEA

The competition for highly accessible sites in the CBD gives rise to a hierarchy of land uses, with a variation in the intensity of land and building use between the innermost part of the CBD - the core - and its surrounding frame.

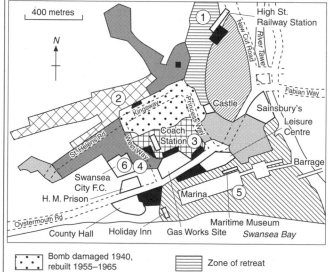

① Zone of retreat - this northern zone is losing businesses because of its distance away from the core.

② Zone of offices - functional adjacency of legal, financial, and estate agents offices; these activities locate near the core for accessibility but do not need the central location required by department stores.

③ The core - this zone includes a pedestrian precinct and a covered shopping mall, the Quadrant Centre.

④ Zone of advance - Swansea is growing westwards, and new buildings and renovation are improving the quality of shops and offices in this area.

⑤ Marina redevelopment - the Marina development is a high status regeneration including houses, shops, a superstore, and museum.

⑥ Frame - activities and land use include transport terminals (the railway in the north), warehousing, and car sales rooms; in Swansea the football club is in the frame.

Legend:
- Bomb damaged 1940, rebuilt 1955–1965
- Redevelopment post-1978
- Marina redeveloped for residential and recreational uses, 1960s
- Former North Dock - redeveloped for leisure and shopping, post–1985
- Older surviving business
- Zone of retreat
- Public buildings sector
- Legal, financial, and estate agents zone
- Zone of advance
- Main car parks (ground and multi-storey)

Inner city

PROCESSES AFFECTING INNER CITIES

- lack of services
- unemployment
- drugs and crime
- urban decay

- high population density
- road congestion and pollution
- lack of green space

Ageing
- relates to housing, social services, infrastructure, industrial base, population
- the result is that population is more dependent on social and medical services; housing is in need of renovation or replacement; transport systems are outmoded; urban infrastructure in need of repair

Changing land uses
- new construction related to the activities of the central business district, new roads, universities, hospitals
- slum clearance associated with tower blocks and moving the inner city poor to the periphery
- urban policy can also change land use with respect to social class as middle class housing replaces working class areas

Changing social structure
- minority groups such as blacks or new immigrants concentrate in parts of the area
- high concentrations of ethnic minorities may be due to internal (defence, support, preservation) or external factors
- population decline and housing markets which 'lock in' the poor
- gentrification may reverse this

CASE STUDY: ETHNICITY - ASIANS IN LEICESTER

Ethnic concentrations are a characteristic of the inner city. Ethnic minorities are groups who are culturally differentiated from the majority population. Visibility is due to language, religion, and race. Concentration in the inner city is due to 'uncontested filtering' into housing made available by the out-migration of the indigenous population (very similar to the Burgess model). In Leicester, immigrants from India and Pakistan and from other British cities were drawn by the following pull factors:

- jobs in industry and textiles and expanding engineering plants

- kinship ties and a resilient economy which drew Asians from other British cities

There were three clear waves of migration:

(i) early post-war male migration (1950-60); the main motive was economic and the move was seen as short term

(ii) family reunion (1960-68); this saw families moving from Asia to join with families, and was a period of house buying and voluntary segregation to retain cultural identity

(iii) East African Asian migration; more middle class than original migrants, and this has led to internal segregation

Map 1

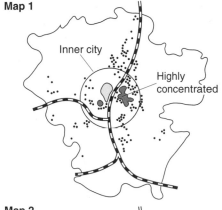

Distribution of Muslims

Muslims are concentrated in Highfields and after 1960 showed a strong voluntary segregation.

Map 2

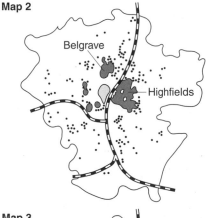

Distribution of Sikhs

Sikhs are concentrated in both Belgrave and Highfields.
They are the group most willing to mix with other Asian communities.

Map 3

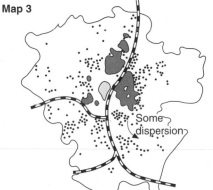

Distribution of Hindus

Some Hindus settled in Highfields, but the East African Asians also chose Belgrave, leading to internal differentiation within the community.

- - - ▪ - - Railway
- ☐ City centre
- ▨ Over 3 households per street
- ·.· Two households per street

FACTORS AFFECTING THE MODERN WESTERN CITY

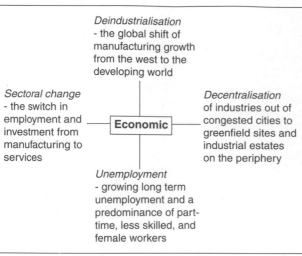

Deindustrialisation
- the global shift of manufacturing growth from the west to the developing world

Sectoral change
- the switch in employment and investment from manufacturing to services

Economic

Decentralisation
of industries out of congested cities to greenfield sites and industrial estates on the periphery

Unemployment
- growing long term unemployment and a predominance of part-time, less skilled, and female workers

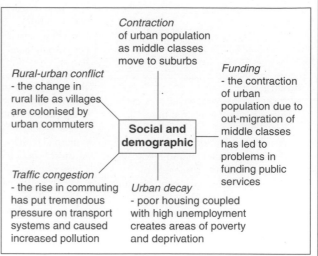

Contraction
of urban population as middle classes move to suburbs

Rural-urban conflict
- the change in rural life as villages are colonised by urban commuters

Funding
- the contraction of urban population due to out-migration of middle classes has led to problems in funding public services

Social and demographic

Traffic congestion
- the rise in commuting has put tremendous pressure on transport systems and caused increased pollution

Urban decay
- poor housing coupled with high unemployment creates areas of poverty and deprivation

CASE STUDY: URBAN REDEVELOPMENT IN GLASGOW

Slum clearance (1957-74)
- comprehensive redevelopment of the tenement areas (Govan, Gorbals, Royston) of the inner city, which were cleared by bulldozers

- existing communities were broken up and forced to relocate to bleak peripheral estates

Peripheral council housing (1952 to the early 1970s)
Problems - post-war housing was high density (700 persons per acre) and poorly maintained, with a lack of basic amenities (over 50% of all houses with no bath) and infested by vermin, especially rats

Policy - 500,000 people were dispersed to new towns like East Kilbride and Cumbernauld and peripheral council housing at Castlemilk (8500), Pollock (9000), Drumchapel (7500), and Easterhouse (10,000)

Evaluation - estates lacked amenities; displaced and divided communities; poor design and rushed construction led to problems of damp and vermin

TOP-DOWN
Government-generated initiatives which attempt to change social or economic structure.

The GEAR project (1976-87)
Problems - Glasgow Eastern Area Renewal (GEAR) was a response to the mistakes of slum clearance and peripheral estates

Policy - modernisation rather than demolition with newly-built housing (2000 private homes) combining with existing housing stock

Evaluation - succeeded in attracting 300 new factories and improving housing (1200 homes were rehabilitated); stopped out-migration of residents; although 'job-rich', most jobs are taken by commuters

Transport
- Glasgow's ambitious transport policy included the construction of one of the UK's few 'urban freeways', the M8

- improved communications led to an increase in commuting and the loss of some good quality inner city housing

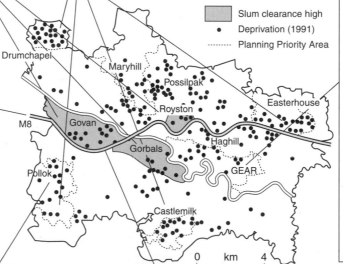

Slum clearance high
• Deprivation (1991)
Planning Priority Area

Drumchapel
Maryhill
Possilpak
Royston
Easterhouse
M8
Govan
Gorbals
Haghill
Pollok
GEAR
Castlemilk

0 km 4

Deprivation
- despite planning policies, multiple deprivation still exists, especially in the peripheral housing schemes

- many argue that long-term unemployment cannot be solved by urban planners

- others argue that newer policies like GEAR ignore social problems

The Govan initiative (1987-94)
Problems - factory closure; decay and decline of housing stock; environmental damage by M8 motorway

Solutions - small-scale developments including start-up units for businesses, environmental improvements (landscaping), education and training for resident workforce

Success/failure - rebirth of local shipyard providing jobs for local workers; 'bottom-up' approach served the needs of the community

BOTTOM-UP
Locally-based initiatives including small-scale social action.

New towns and green belts

CASE STUDY: THE GREATER LONDON PLAN

Until the mid-1960s the basis for planning in the South-East was the 1944 Greater London Plan ('The Abercrombie Plan'). The plan was set up to solve a number of problems:

- London was too large. Too many of the UK's jobs were centred on London.
- Other areas were suffering unemployment. London was congested. Many dwellings were slums.

New towns

New Towns Act, 1946

- eight New Towns were created around London
  - with target populations of between 25,000 and 80,000

- 28 settlements were expanded to take another 535,000 migrating Londoners ('expanded towns')

- New Towns were set up to provide alternatives to London in terms of housing and employment

Green belt

Green Belt Act, 1938

- a zone of land around London within which building is controlled

- 25 km wide; has many towns within it, which can only expand by infilling the spaces between existing buildings

- set up to stop the sprawl of London and the merging of neighbouring towns, to protect farmland, and to restrict harmful activities on rural-urban fringe

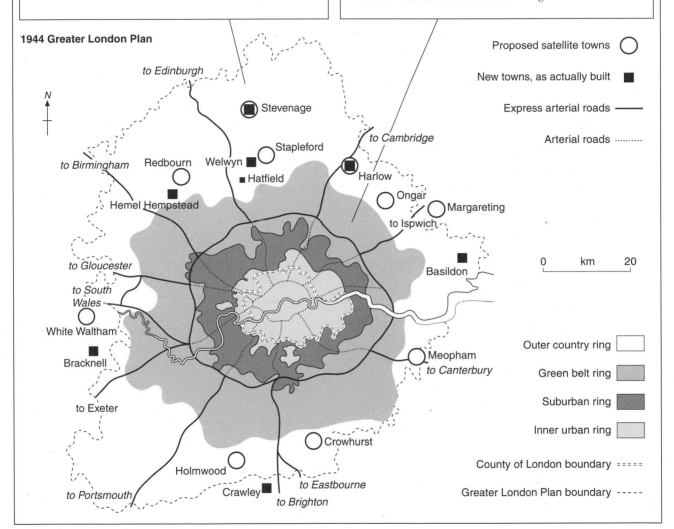

1944 Greater London Plan

Proposed satellite towns ○
New towns, as actually built ■
Express arterial roads ———
Arterial roads ----------

Outer country ring
Green belt ring
Suburban ring
Inner urban ring
County of London boundary =====
Greater London Plan boundary -----

NEW TOWNS - SUCCESS OR FAILURE?

- First wave were close enough to London to allow daily commuting
  - they helped to relieve the housing problem
  - but added to congestion

- Second wave were built further out and have become autonomous growth poles, e.g. Milton Keynes.

GREEN BELT - SUCCESS OR FAILURE?

- Has succeeded in protecting many rural areas from urban sprawl.

- Problems include:
  - increased commuting as urban dwellers relocate outside the green belt
  - house prices in London forced up
  - many argue that much of the green belt is poor quality and not worth preserving
  - market forces have seen planning regulations relaxed, e.g. the construction of the M25

The developing world city

A country's development is often measured in terms of how urbanised it is. This has been questioned. Regarding cities in the developing world, arguments are based on whether city functions generate developments or whether they create problems.

FUNCTIONS

- *Commercially* towns provide the market and exchange centres necessary for the conversion from subsistence to cash crops.

- *Industrially* towns provide a focus for investment and labour.

- *Politically* urbanisation generates a national outlook rather than regional or tribal insularity.

- *Administratively* cities provide a centre for health and education.

- *Internationally* cities provide the business infrastructure needed for western investment.

PROBLEMS

- Over-urbanisation results in housing shortages, traffic congestion, pollution, over-stretched services, and unemployment and underemployment.

- Primate cities (more than twice the size of the next largest) can suck investment from rural areas, adding to the problems of rural-urban migration and damaging national development.

- Rapid growth through migration and natural increase means housing conditions are poor with many residents faced with a lack of sanitation and piped water, illness, and crime.

JFC TURNER'S MODEL (1970)

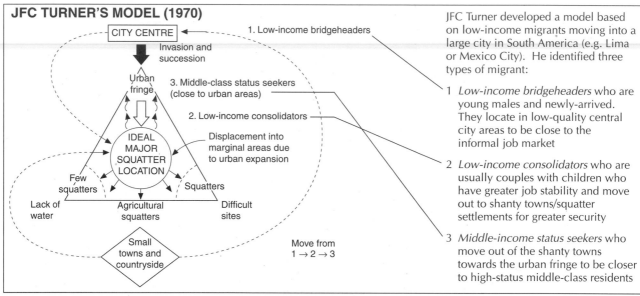

JFC Turner developed a model based on low-income migrants moving into a large city in South America (e.g. Lima or Mexico City). He identified three types of migrant:

1. *Low-income bridgeheaders* who are young males and newly-arrived. They locate in low-quality central city areas to be close to the informal job market

2. *Low-income consolidators* who are usually couples with children who have greater job stability and move out to shanty towns/squatter settlements for greater security

3. *Middle-income status seekers* who move out of the shanty towns towards the urban fringe to be closer to high-status middle-class residents

THE URBAN STRUCTURE OF MEXICO CITY

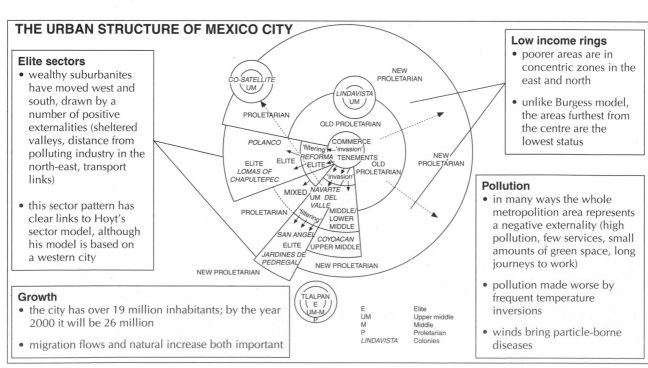

Elite sectors
- wealthy suburbanites have moved west and south, drawn by a number of positive externalities (sheltered valleys, distance from polluting industry in the north-east, transport links)

- this sector pattern has clear links to Hoyt's sector model, although his model is based on a western city

Growth
- the city has over 19 million inhabitants; by the year 2000 it will be 26 million

- migration flows and natural increase both important

Low income rings
- poorer areas are in concentric zones in the east and north

- unlike Burgess model, the areas furthest from the centre are the lowest status

Pollution
- in many ways the whole metropolition area represents a negative externality (high pollution, few services, small amounts of green space, long journeys to work)

- pollution made worse by frequent temperature inversions

- winds bring particle-borne diseases

E	Elite
UM	Upper middle
M	Middle
P	Proletarian
LINDAVISTA	Colonies

Agricultural systems

Agriculture is the harvesting of crops and animal products for human and/or animal consumption and for industrial production.

CLASSIFYING AGRICULTURE

The following are not exclusive categories but indicate a scale along which all farming types can be placed.

Arable: the cultivation of crops, e.g. wheat farming in East Anglia.
Pastoral: the rearing of animals, e.g. sheep farming in the Lake District.

Commercial: products are sold to make a profit, e.g. market gardening in the Netherlands.
Subsistence (or **peasant farming**): products are consumed by the cultivators, e.g. shifting cultivation by the Kayapo indians in the Amazonian rainforest.

Intensive: high inputs or yields per unit area, e.g. battery hen production.
Extensive: low inputs or yields per unit area, e.g. free range chicken production.

Nomadic: farmers move seasonally with their herds, e.g. the Pokot, pastoralists in Kenya.
Sedentary: farmers remain in the same place throughout the year, e.g. dairy farming in Devon and Cornwall.

FACTORS AFFECTING AGRICULTURE

Physical factors

Climate	Precipitation	• type
		• frequency
		• intensity
		• amount
	Temperature	• growing season (> 6°C)
		• ground frozen (0°C)
		• range of temperatures
Soil	Fertility	• pH
		• cation exchange capacity
		• nutrient status
	Structure	
	Texture	
	Depth	
Pests	Vermin, locusts, disease etc.	
Slope	Gradient	
Relief	Altitude	
Aspect	Ubac (shady) or adret (sunny)	

Human factors

Political	Land tenure/ownership	• ownership, rental, share-cropping, state-control
	Organisation	• collective, co-operative agribusiness, family farm
	Government policies	• subsidies, guaranteed prices, ESAs, quotas, set-aside
	War	• disease, famine
Economic	Farm size	• field size and shape
	Demand	• size and type of market
	Capital	• equipment, machinery, seeds, money, 'inputs'
	Technology	• HYVs, fertilisers, irrigation
	Infrastructure	• roads, communications, storage
	Advertising	
Social	Cultural and traditional influences	
	Education and training	
	Behavioural influences	
	Chance	

Farming systems

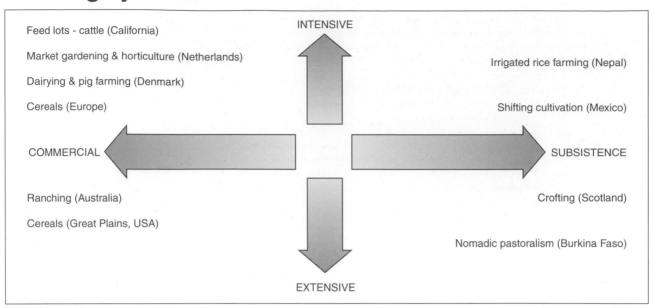

Feed lots - cattle (California)

Market gardening & horticulture (Netherlands)

Dairying & pig farming (Denmark)

Cereals (Europe)

COMMERCIAL

Ranching (Australia)

Cereals (Great Plains, USA)

INTENSIVE

Irrigated rice farming (Nepal)

Shifting cultivation (Mexico)

SUBSISTENCE

Crofting (Scotland)

Nomadic pastoralism (Burkina Faso)

EXTENSIVE

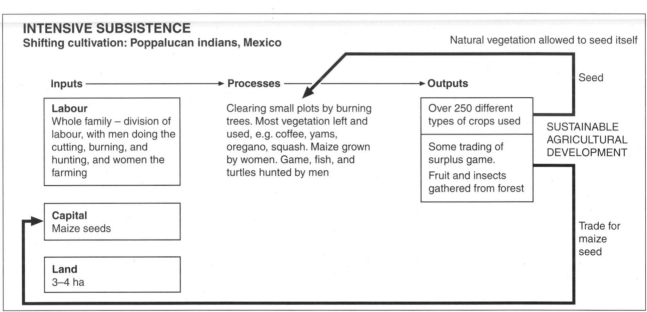

INTENSIVE SUBSISTENCE
Shifting cultivation: Poppalucan indians, Mexico

Natural vegetation allowed to seed itself

Inputs ⟶ Processes ⟶ Outputs

Seed

Labour
Whole family – division of labour, with men doing the cutting, burning, and hunting, and women the farming

Clearing small plots by burning trees. Most vegetation left and used, e.g. coffee, yams, oregano, squash. Maize grown by women. Game, fish, and turtles hunted by men

Over 250 different types of crops used

Some trading of surplus game.

Fruit and insects gathered from forest

SUSTAINABLE AGRICULTURAL DEVELOPMENT

Capital
Maize seeds

Land
3–4 ha

Trade for maize seed

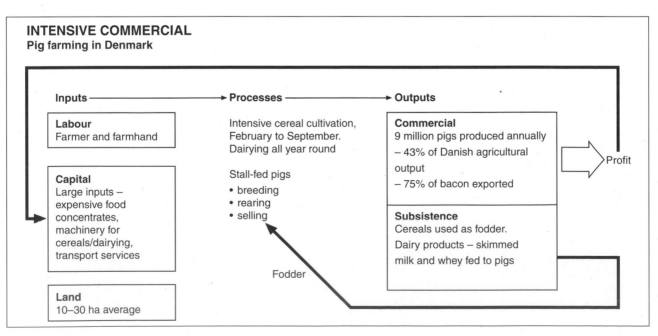

INTENSIVE COMMERCIAL
Pig farming in Denmark

Inputs ⟶ Processes ⟶ Outputs

Labour
Farmer and farmhand

Capital
Large inputs – expensive food concentrates, machinery for cereals/dairying, transport services

Land
10–30 ha average

Intensive cereal cultivation, February to September. Dairying all year round

Stall-fed pigs
• breeding
• rearing
• selling

Fodder

Commercial
9 million pigs produced annually
– 43% of Danish agricultural output
– 75% of bacon exported

Subsistence
Cereals used as fodder.
Dairy products – skimmed milk and whey fed to pigs

Profit

Agricultural ecosystems

Agricultural ecosystems can be compared with natural ecosystems in terms of productivity, biomass, nutrient cycling, and energy efficiency. Average productivity from agricultural systems is 650 g/m^2/yr, comparable with temperate grasslands or prairies. However, only about 0.25% of incoming radiation is utilised by crops, and of that less than 1% is harnessed by people through food.

ECOSYSTEMS

	Natural ecosystems	Agricultural ecosystems
Foodweb	Complex; several layers	Simple; mostly one or two layers
Biomass	Large; mixed plant and animal	Small; mostly plant
Biodiversity	High	Low - often mono-culture
Gene pool	High	Low, e.g. three species of cotton account for 53% of crop
Nutrient cycling	Slow; self-contained; unaffected by external supplies	Largely supported by external supplies
Productivity	High	Lower
Modification	Limited	Extensive - inputs of feed, seed, water, fertilisers, energy fuel; outputs of products, waste

EFFECTS OF NOMADIC PASTORALISM

1. **Energy flow**
   Little change - cattle replace wild herbivores; limited killing of predators.

2. **Nutrient cycling**
   Little change - some concentrations of nutrients, in dung, may occur if herds remain in one place for a length of time. True nomadic movement returns and distributes nutrients over a wide area.

3. **Biological productivity**
   NPP low and variable - 150 g/m^2/yr in drier areas rising to 600 g/m^2/yr in wetter margins. Secondary productivity is low - hence farmers use milk, milk products, and blood, rather than meat.

4. **Ecological stability and modification**
   Over-exploitation of grass or over-concentration of herds removes vegetation, especially sweeter species, causing ponding of the surface, gullying, and desertification - the spread of desert conditions. This was due to climatic deterioration in traditional pastoral societies, but is increasingly common for economic, social, and political reasons: larger herds, shorter nomadic routes, and greater pressure around water sources, e.g. boreholes.

UK NUTRIENT CYCLES

Temperate deciduous woodland

Input dissolved in rain

Leaf fall, tissue decay

Run-off

B

L

Mineralisation, humification, and degradation

S

Uptake by plants

Weathering of rocks

Mixed farming

Harvesting crops, livestock manure

B

L

S

Legumes

Fertilisers

B Biomass
L Litter
S Soil

ENERGY RATIOS (ER)

$$ER = \frac{\text{Energy outputs}}{\text{Energy inputs}}$$

Shifting cultivation	65.0
Hunter-gatherers	7.8
UK cereal farm	1.9
UK allotment	1.3
UK dairy farm	0.38
Broiler hens	0.1
Greenhouse lettuces	0.002

The Common Agricultural Policy (CAP)

The **Common Agricultural Policy** was developed to achieve four main goals:
1. To increase agricultural **productivity** and **self-sufficiency**.
2. To ensure a **fair standard of living** for farmers.
3. To **stabilise markets**.
4. To ensure food was available to consumers at a **fair price**.

Three principles lay behind this:
1. A single European agricultural market in which goods could move freely.
2. Preferential treatment for European food.
3. EEC (now the EU) funding of the CAP.

The CAP achieved its main aim of increasing food supply by **guaranteed prices** and **intervention buying and storage**, i.e. a guaranteed market. This led to **intensification**, **specialisation**, and **concentration** of agricultural activities in the better suited areas.

CAP REFORM

The CAP created huge surpluses - by the early 1990s, 33 million tons of cereal, 2200 million litres of wine, and 8 million tonnes of beef had to be stored. The cost of storing butter alone was over £5 million per week.

In 1992, the CAP was reformed. Surplus production was inefficient and costly to store, subsidies were excessively large, and intensive farming was harming the environment. Changes included reduction of price support, increased quotas, extensification of agriculture, and set-aside. However, by 1995-6, owing to the reforms of the CAP and a series of very hot summers, agricultural surpluses had been drastically reduced.

The worst food crisis since 1974 has left the world with only 53 days supply of grain. The world is seriously short of food and up to 35,000 children die from hunger related diseases every day. Worldwide grain stocks are well below the FAO's minimum necessary to safeguard world food security. In 1987 there was over 100 days' worth of food supplies, but in 1995 there were just over 50, and by the end of 1996 it will be below 50 days'.

World food production has lagged behind food consumption since 1993. The drought of 1995 has led to the lowest harvest of food/head since the mid-1970s. The blistering summer of 1995, the hottest recorded in many parts of the northern hemisphere, destroyed millions of crops. The drought in Spain entered its fourth year and wheat yields slumped to less than half of their 1994 levels.

The European food mountain is fast being eroded. Its grain mountain has fallen from 33 million tons in 1993 to just 5.5 million tons in 1995. The only significant surplus is the wine lake, at 120 million litres.

Europe is not producing enough food, partly as a result of the 1992 CAP reforms which increased the amount of set-aside. Plans to reduce set-aside to 10% of land, rather than 15% in 1995, are an attempt to increase food production.

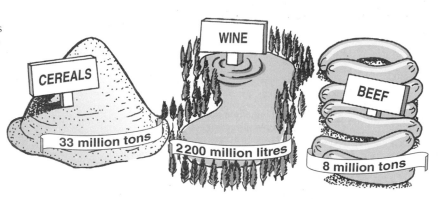

CEREALS — 33 million tons

WINE — 2200 million litres

BEEF — 8 million tons

HOW THINGS WILL CHANGE

Under the reforms proposed by the EU in 1992, the form of support for farmers changed. The price Brussels offered to farmers fell, but farmers had their income topped up by around £188 per hectare, provided they took 15% of their land out of production under a scheme known as 'set-aside'. The result was more big farms, more golf courses, and more uncultivated land.

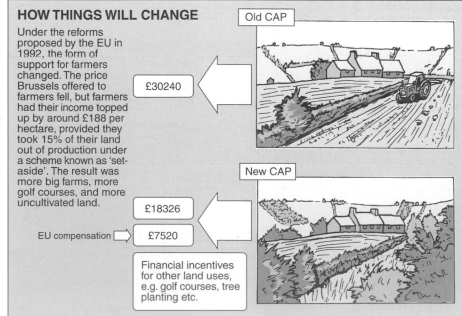

Old CAP

£30240

New CAP

£18326

EU compensation — £7520

Financial incentives for other land uses, e.g. golf courses, tree planting etc.

West Midgeley Farm
40-hectare cereal farm

Assuming a crop yield of 7 tonnes a hectare, West Midgeley Farm will produce 280 tonnes. At the current price of £108 a tonne, the farmer would have an income of £30240.

West Midgeley Farm
40-hectare cereal farm

Under EU reforms, the farmer has to reduce the area by 15%, leaving 34 hectares. The price that would be guaranteed from the crops would be £77 a tonne, giving an income of £18326. But the farmer's income would be topped up by compensation of £188 a hectare, giving an extra £7520. Total: £18326 plus £7520 equals £25846. On top of this, however, the EU has said there will be other financial incentives.

The green revolution

THE PROBLEM

Population growth is more rapid than the increase in food production. In India, for example, by 2000 AD the population will reach 1 billion people and food production will need to increase by 40% to match demand. But much of India's land is of limited potential.

THE SOLUTION?

The **green revolution** is the application of science and technology to increase crop productivity. It includes a variety of techniques such as genetic engineering to produce higher yielding varieties (HYVs) of crops and animals, mechanisation, pesticides, herbicides, chemical fertilisers, and irrigation water.

HYVs are the flagship of the green revolution. During 1967-8 India adopted Mexican Rice IR8 which had a short stalk and a larger head than traditional varieties, and yielded twice as much grain. However, it required considerable amounts of water and nitrogen. Up to 55% of India's crops are now HYVs. 85% of the Philippines' crops are HYVs; by contrast only 13% of Thailand's crops are HYVs.

ADOPTION OF HYVs

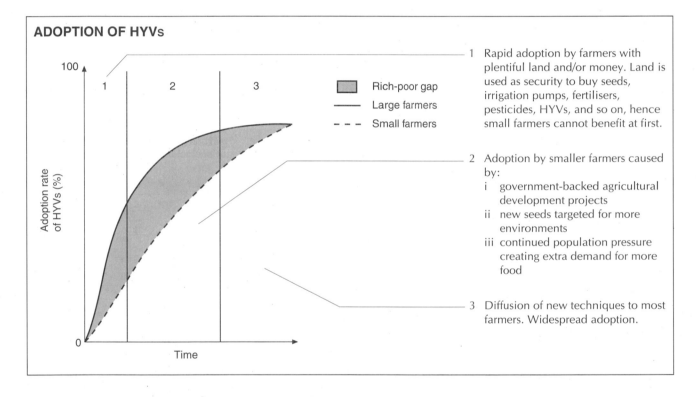

Rich-poor gap
Large farmers
Small farmers

1 Rapid adoption by farmers with plentiful land and/or money. Land is used as security to buy seeds, irrigation pumps, fertilisers, pesticides, HYVs, and so on, hence small farmers cannot benefit at first.

2 Adoption by smaller farmers caused by:
   i   government-backed agricultural development projects
   ii new seeds targeted for more environments
   iii continued population pressure creating extra demand for more food

3 Diffusion of new techniques to most farmers. Widespread adoption.

THE CONSEQUENCES

The main benefit is that more food can be produced:
- yields are higher
- up to three crops can be grown each year
- more food should lead to less hunger
- more exports create more foreign currency

However, there are many problems:
- not all farmers adopt HYVs - some cannot afford the cost
- as the cost rises, indebtedness increases
- rural unemployment has increased due to mechanisation
- irrigation has led to salinisation - 20% of Pakistan's and 25% of Central Asia's irrigated land is affected
- soil fertility is declining as HYVs use up all the nutrients; these can be replenished by fertilisers, but this is expensive
- LDCs are dependent on many developed countries for the inputs

Changes in South India: the effects of the green revolution	
Use of fertiliser	+138%
Human labour	+111%
Paddy rice	+91%
Sugar cane	+41%
Income	+20%
Subsistence food	-90%
Energy efficiency	-25%

Agriculture and environmental issues in the UK

SOIL EROSION

- Over one-third of arable land in the UK is at risk of soil erosion.
- Sandy and sandy-loam soils with a slope angle of more than 3° are particularly vulnerable.
- Soil losses are up to 250 t/ha on the South Downs, 160 t/ha in Norfolk, and 150 t/ha in West Sussex.

The potential for soil erosion has increased considerably in recent years for a number of reasons:

1 Spread of arable land use into pastoral areas.
2 Hedgerow removal.
3 Ploughing and draining of peaty soils.
4 Afforestation leaves bare ground between young trees.
5 Increased recreational pressure in rural areas.

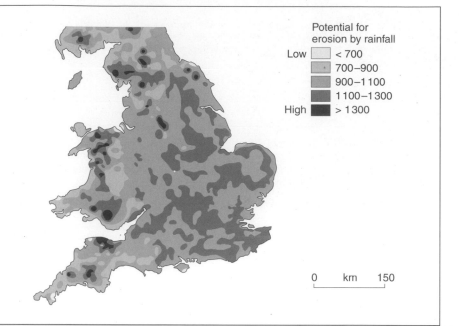

Potential for erosion by rainfall

Low		< 700
		700–900
		900–1100
		1100–1300
High		> 1300

0   km   150

NITRATE POLLUTION

The use of nitrate fertilisers increased from 200,000 tonnes in 1945 to a peak of 1.6 million tonnes in the late 1980s.

Increased levels of nitrates in ponds and streams can cause **eutrophication**, i.e. nutrient enrichment. This can lead to algal blooms, which can cause a shortage of oxygen and light and thus declining biodiversity.

Nitrate levels in **drinking water** continue to rise with many rivers exceeding the EU safe limit of 11.3 mg/l. Up to 5 million people in England and Wales are supplied with water with high rates of nitrates. This is linked with:

- high rates of stomach cancer
- blue baby syndrome, due to oxygen starvation in the bloodstream

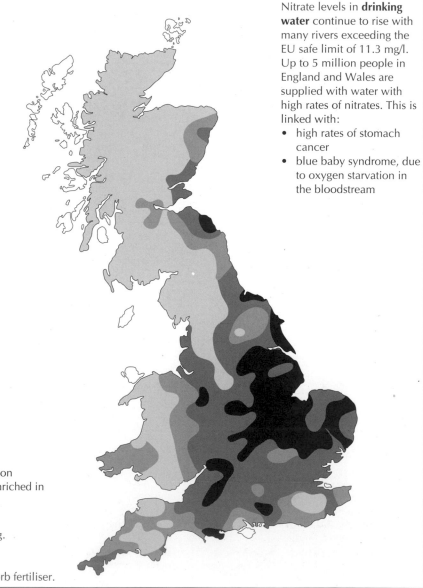

$NO_3 - N$ concentration (mg/l^{-1})

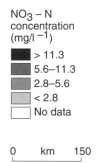

	> 11.3
	5.6–11.3
	2.8–5.6
	< 2.8
	No data

0   km   150

It costs between £50 million and £300 million annually to purify water that has become enriched in nitrates! Prevention is cheaper than cure.

Solutions include:

1 A change in land use to pastoral farming.
2 Less intensive arable agriculture.
3 Less use of fertiliser.
4 The use of cover crops in winter to absorb fertiliser.

Reducing the environmental effects of agriculture

SET-ASIDE

The **set-aside** scheme was introduced on a voluntary basis in 1988 allowing farmers to take up to 20% of their land out of production and to receive up to £200 for each hectare set aside. The land could be left fallow, converted to woodland, or used for non-agricultural purposes. Reform of the CAP in 1992 reduced the amount of set-aside to a maximum of 15%, and further reform in 1994 reduced rotational set-aside to 12% and flexible set-aside to 15%. While many farmers took advantage of set-aside, many intensified production on the other land and made their least favourable land the set-aside! Between 1992 and 1993 the total area in England and Wales under cereals decreased by 400,000 hectares and there was a similar increase in the amount of set-aside.

ESAs

In 1985 the EU agreed to provide farmers with the means to farm **environmentally sensitive areas** in traditional ways which would preserve important biological and heritage landscapes. Less intensive, organic methods were favoured with increased amounts of fallow. By 1994, 10,500 farmers had signed or applied for ESA agreements, and payments during 1994/5 totalled about £25 million.

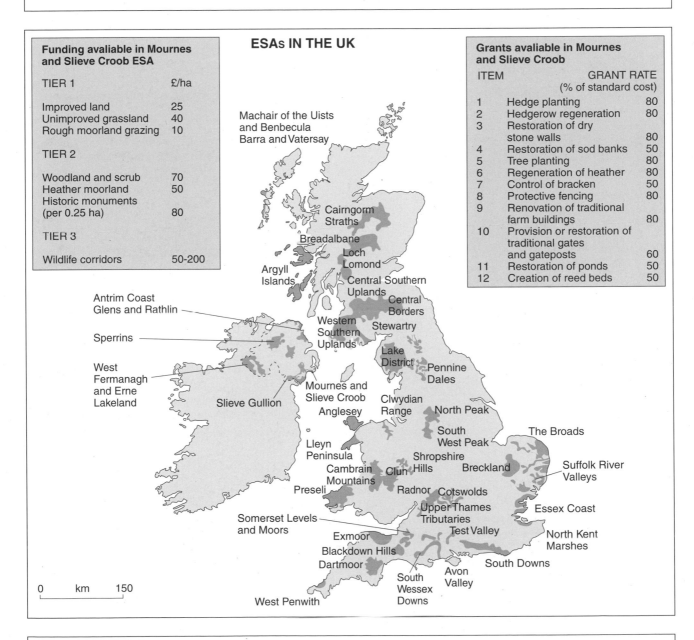

ESAs IN THE UK

Funding avaliable in Mournes and Slieve Croob ESA

TIER 1	£/ha
Improved land	25
Unimproved grassland	40
Rough moorland grazing	10

TIER 2	
Woodland and scrub	70
Heather moorland	50
Historic monuments (per 0.25 ha)	80

TIER 3	
Wildlife corridors	50-200

Grants avaliable in Mournes and Slieve Croob

ITEM		GRANT RATE (% of standard cost)
1	Hedge planting	80
2	Hedgerow regeneration	80
3	Restoration of dry stone walls	80
4	Restoration of sod banks	50
5	Tree planting	80
6	Regeneration of heather	80
7	Control of bracken	50
8	Protective fencing	80
9	Renovation of traditional farm buildings	80
10	Provision or restoration of traditional gates and gateposts	60
11	Restoration of ponds	50
12	Creation of reed beds	50

Other schemes include:
- **Nitrate Sensitive Areas** to protect groundwater areas.
- **Habitat Schemes** to improve/create wildlife habitats.
- **Organic Aid Schemes** to encourage farmers to convert to organic production methods.
- The **Countryside Access Scheme** to grant new opportunities for public access to set-aside land and suitable farmland in ESAs.

Agricultural models

VON THUNEN'S MODEL

Johann Von Thunen's model of **locational or economic rent** (1826) suggests that land use and intensity of production declines with distance from a central market. High intensity market gardening, dairying, and horticulture predominate close to urban areas while extensive grain and livestock farming are located furthest away. Woodland was an important land use when Von Thunen developed his model and was found close to the urban area. Although his model is criticised for its simplicity and its assumptions (that farmers' sole aim is to maximise profits, i.e. **rational man**, and that physical conditions do not vary, i.e. an **isotropic plain**), aspects of his model can be observed at a variety of scales, from the individual farm up to land use in Europe. It is also important whenever transport is poorly developed, especially in developing countries.

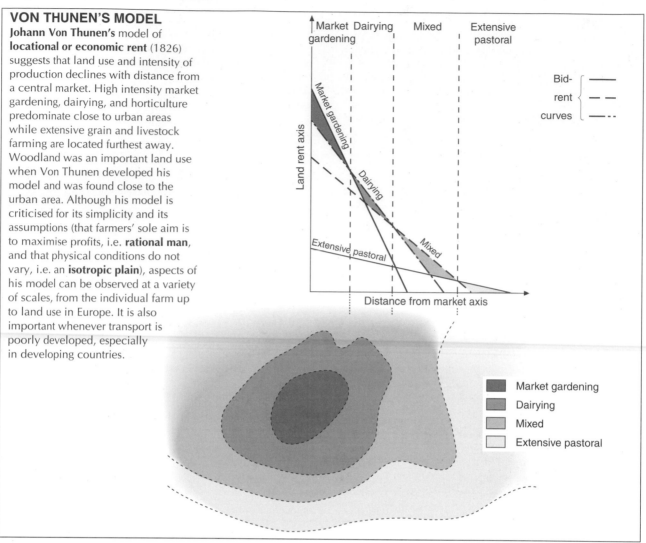

SINCLAIR'S MODEL

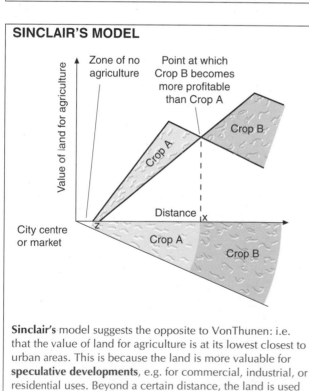

Sinclair's model suggests the opposite to VonThunen: i.e. that the value of land for agriculture is at its lowest closest to urban areas. This is because the land is more valuable for **speculative developments**, e.g. for commercial, industrial, or residential uses. Beyond a certain distance, the land is used for agriculture as it loses its value for development.

HAGERSTRAND'S MODEL

Hagerstrand showed how **new innovations** and **techniques** were likely to be used only by a few people at first (innovators) before being adopted rapidly, although a few laggards would resist change. This meant the adoption of any technique followed an S-shaped curve.

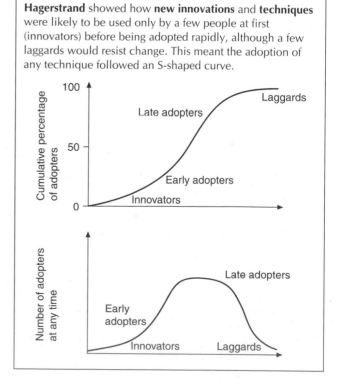

Industrial location

FACTORS INFLUENCING LOCATION

- Physical factors influenced industrial location in the nineteenth century.

- Old industrial regions were located where cheap energy and raw materials were found.

- Physical factors influenced mature industries like textiles, shipbuilding, and iron and steel.

- Transport costs and markets also influenced mature industries.

- In the 1990s, government policy and labour requirements have strong influences on industry.

- Industrial location is no longer country-specific and choices of location are global and strategic.

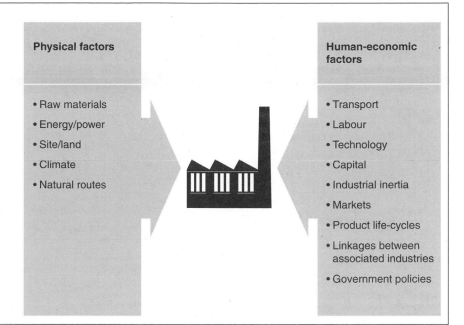

Physical factors

- Raw materials
- Energy/power
- Site/land
- Climate
- Natural routes

Human-economic factors

- Transport
- Labour
- Technology
- Capital
- Industrial inertia
- Markets
- Product life-cycles
- Linkages between associated industries
- Government policies

CLASSIFICATION

Primary industries The extraction of raw materials, e.g. mining, quarrying, farming, fishing, and forestry.

Secondary industries Manufacturing industries which involve the transformation of raw materials (or components) into consumable products, e.g. steelworks, the car industry, and high technology.

Tertiary industries These are concerned with providing a service to customers, e.g. transport, retailing, and medical and professional services.

Quaternary industries These provide information and expertise, e.g. universities, research and development, media, and political policy units.

THE INDUSTRIAL SYSTEM

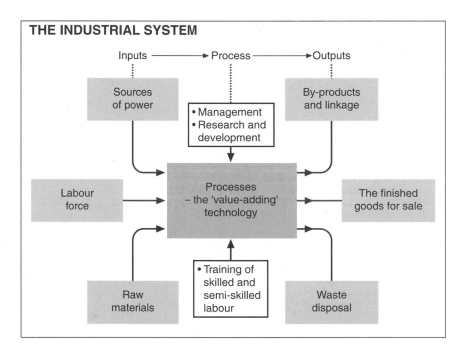

Agglomeration economies Savings which arise from the concentration of industries either together or close to linked activities.

Industrial inertia The survival of an industry in an area even though the initial advantages are no longer relevant.

Break-of-bulk location A location which takes its advantage from a position where there is forced transfer of freight from one transport medium to another, e.g. a port or rail terminal.

Greenfield site An industrial site located on the edge of an urban area in a place with no prior industrial use.

DEFINITIONS OF KEY WORDS

Rationalisation A reduction in the production capacity of a multi-plant firm by factory closure.

Transplant An assembly plant owned and operated by a foreign-based company.

Transnational corporation A large, multi-plant firm with a worldwide manufacturing capability.

Research and development (R&D) The branch of a manufacturing firm concerned with the design and development of new products; R&D employs highly skilled workers.

Classical location theory

All models of industrial location simplify the real world in order to illustrate one or more concepts which influence locational choice. There are four traditional location models.

1 WEBER'S LEAST-COST LOCATION MODEL (1926)

Weber concentrated on *costs* to explain the *optimum location* for industry.

(i) Transport costs

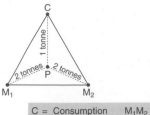

C =	Consumption point (market)	M_1M_2 =	Raw materials
		P =	Optimum location

a. Weight-losing industry

Raw materials are heavier than finished product so transport costs are saved by locating close to the raw materials

Example: Iron and steel, which uses bulky raw materials

b. Weight-gaining industry

Finished product is heavier than raw materials so location is closer to the market

Example: Brewing, which gains weight through the addition of a ubiquitous raw material, water

The optimum location may shift if savings from labour costs or agglomeration outweigh the increased transport costs of moving.

(ii) Labour costs

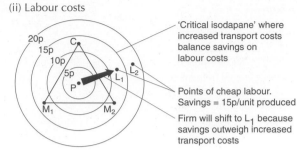

'Critical isodapane' where increased transport costs balance savings on labour costs

Points of cheap labour. Savings = 15p/unit produced

Firm will shift to L_1 because savings outweigh increased transport costs

(iii) Agglomeration forces

Some firms can save money by locating together. This could be by sharing costs (industrial estates) or ideas (science parks).

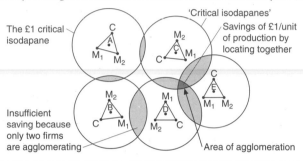

The £1 critical isodapane

'Critical isodapanes'

Savings of £1/unit of production by locating together

Insufficient saving because only two firms are agglomerating

Area of agglomeration

2 LÖSCH'S PROFIT MAXIMISATION MODEL (1954)

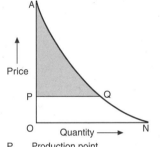

P	Production point
OP	Price at production point
AQN	Demand curve
PQ	Quantity sold at P
A	No demand because price too high

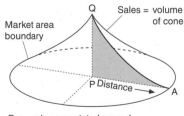

Demand curve rotated around production point to give cone AQP

Lösch's model is based on weight-gaining industries like brewing and baking. It states that sales will fall with increasing distance from production due to higher transport costs. Unlike Weber, *optimum location* is defined by maximum profit rather than minimum costs.

3 PRED'S BEHAVIOURAL MATRIX (1967)

Pred's model relaxes Weber's assumption of 'economic man' acting with perfect knowledge. Sub-optimal locations are based on poor knowledge or lack of entrepreneurial flair.

(a) Behavioural matrix

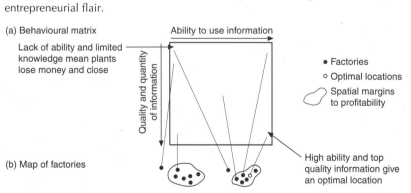

Ability to use information

Lack of ability and limited knowledge mean plants lose money and close

• Factories
○ Optimal locations
⬭ Spatial margins to profitability

Quality and quantity of information

(b) Map of factories

High ability and top quality information give an optimal location

4 THE SMITH AND RAWSTRON MODEL OF SPATIAL MARGINS (1966)

Smith and Rawstron also described sub-optimal location. Their *spatial margins* define whole areas of profitability. The model introduces the idea that firms may not choose the optimum location but rather be content with a satisfactory profit. It also focuses attention on the limits to freedom of locators.

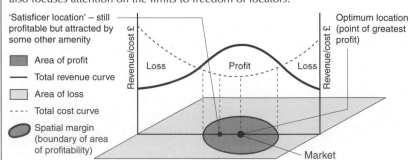

'Satisficer location' – still profitable but attracted by some other amenity

Optimum location (point of greatest profit)

▓ Area of profit
— Total revenue curve
░ Area of loss
---- Total cost curve
⬭ Spatial margin (boundary of area of profitability)

Market

New location theory

Over the last thirty years the influences on industrial location described by the classical models have changed in importance. In the 1990s, *labour* is the most important factor in the location of manufacturing industries. The term *spatial division of labour* is used to describe the way workforce can influence location on a national or global scale. Two models have been suggested.

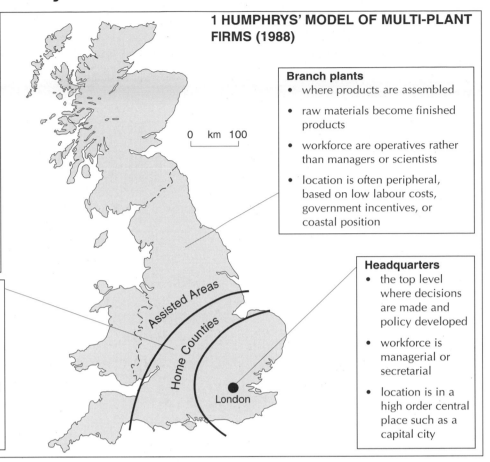

1 HUMPHRYS' MODEL OF MULTI-PLANT FIRMS (1988)

Branch plants
- where products are assembled
- raw materials become finished products
- workforce are operatives rather than managers or scientists
- location is often peripheral, based on low labour costs, government incentives, or coastal position

Headquarters
- the top level where decisions are made and policy developed
- workforce is managerial or secretarial
- location is in a high order central place such as a capital city

Research and development
- new products are developed
- highly qualified technical staff, often with second degrees
- located in core regions and increasingly in semi-rural science parks and technopoles

2 VERNON'S PRODUCT LIFE-CYCLE MODEL (1966)

The product life-cycle model is associated with four stages of production which have clear implications both to the spatial division of labour and industrial location.

Manufacturing capacity shifts from core areas to peripheral areas and finally out of the country as products move through their life cycle.

1 *Infancy* - product is developed and is sold as a niche product to the core region; labour force is skilled but small.

2 *Growth* - demand expands and production techniques become standardised; labour force grows.

3 *Maturity* - little new research and development; production moves to peripheral areas where labour is cheaper.

4 *Obsolescence* - declining sales due to better or cheaper products from abroad; production moves to firms overseas or the product is superseded by new innovation.

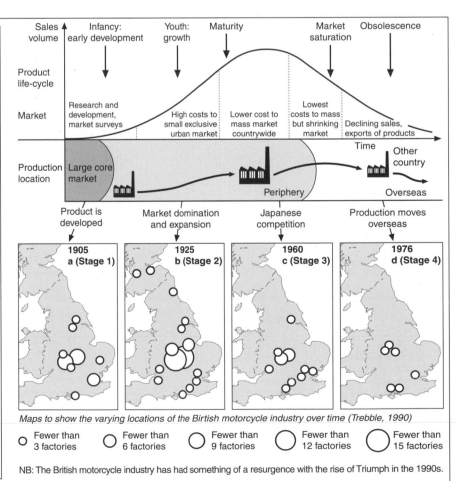

Maps to show the varying locations of the Birtish motorcycle industry over time (Trebble, 1990)

- ○ Fewer than 3 factories
- ○ Fewer than 6 factories
- ○ Fewer than 9 factories
- ○ Fewer than 12 factories
- ◯ Fewer than 15 factories

NB: The British motorcycle industry has had something of a resurgence with the rise of Triumph in the 1990s.

Themes in manufacturing

A number of themes are influencing the location and distribution of manufacturing industry in the 1990s. These include rationalisation and restructuring, internationalisation, just-in-time and flexible production, environmental issues, and urban-rural shift.

1 RATIONALISATION AND RESTRUCTURING

Rationalisation and restructuring involves the drive to improve *competitiveness* and *productivity*. This often means reducing workforces, closing inefficient plants, and changing the way products are made. Unilever's manufacturing base was rationalised from 13 factories down to four between 1973 and 1989.

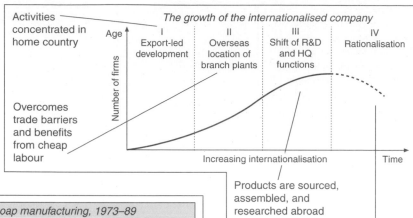

The growth of the internationalised company

Activities concentrated in home country

Overcomes trade barriers and benefits from cheap labour

Age

Number of firms

I Export-led development

II Overseas location of branch plants

III Shift of R&D and HQ functions

IV Rationalisation

Increasing internationalisation

Time

Products are sourced, assembled, and researched abroad

Concentration of activities in the best locations; some profitable plants may be closed down

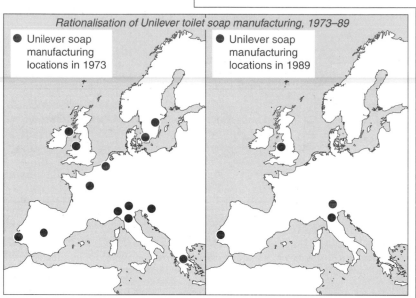

Rationalisation of Unilever toilet soap manufacturing, 1973–89

● Unilever soap manufacturing locations in 1973

● Unilever soap manufacturing locations in 1989

2 INTERNATIONALISATION AND GLOBALISATION

Internationalisation represents the global expansion of a company. It starts from a point where a company begins to export from its home base, through location overseas, to a point where a company finally decides to rationalise its global location in areas of comparative, competitive, or strategic advantage.

3 JUST-IN-TIME (JIT) AND FLEXIBLE PRODUCTION

The essence of JIT is a reduction of stored parts by arranging the provision of parts when they are needed to go into a 'parent item', with delivery on the same day or even every hour. This suggests a shift from Fordist (mass production) to Toyotist (lean production) methods. Lean production is more flexible.

4 ENVIRONMENTAL ISSUES

During the past two decades environmental issues have become an increasingly important part of industrial decision-making. One result is that polluting industries migrate to peripheral or developing world locations.

5 URBAN-RURAL SHIFT

There has been an urban-rural shift of industry over the last 30 years. A number of explanations have been suggested:

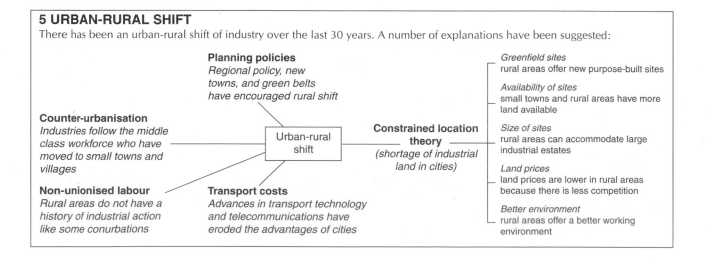

Planning policies
Regional policy, new towns, and green belts have encouraged rural shift

Counter-urbanisation
Industries follow the middle class workforce who have moved to small towns and villages

Urban-rural shift

Non-unionised labour
Rural areas do not have a history of industrial action like some conurbations

Transport costs
Advances in transport technology and telecommunications have eroded the advantages of cities

Constrained location theory
(shortage of industrial land in cities)

Greenfield sites
rural areas offer new purpose-built sites

Availability of sites
small towns and rural areas have more land available

Size of sites
rural areas can accommodate large industrial estates

Land prices
land prices are lower in rural areas because there is less competition

Better environment
rural areas offer a better working environment

Deindustrialisation

In the period since the Second World War the UK has entered a new industrial phase. The characteristics of this so-called *post-industrial society* are:

- a shift away from agriculture and manufacturing industries towards service industries

- the growing importance of large, multinational corporations

- a spatial division of labour within the growing economies of South-East Asia

- the decline of the older 'smokestack' industries

One of the results of post-industrial change has been *deindustrialisation*.

Deindustrialisation is the long-term absolute decline in the manufacturing sector with respect to jobs and production.

The more mature industries like textiles, iron and steel, and shipbuilding are most likely to deindustrialise. The results can be plant closure, job losses, and regional decline.

TYPES OF DEINDUSTRIALISATION

There are two types of deindustrialisation:

Negative deindustrialisation

Plant closure due to inefficiency or obsolescence

⬇

Labour is displaced and high unemployment results

⬇

Services and active population migrate to more prosperous areas

⬇

Regional problems

Positive deindustrialisation

Industries reduce plants and workforce to improve competitiveness

⬇

Productivity is improved

⬇

Displaced labour absorbed by the service sector or new manufacturing firms

⬇

Regional prosperity

EXPLANATIONS OF DEINDUSTRIALISATION

1 Maturity

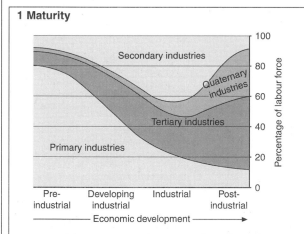

An inevitable progression from agriculture through manufacturing industry to the service industry, i.e. post-industrial change.

3 Management slow to innovate

In the boom period of the 1960s even inefficient plants could make a profit. This meant managers were unwilling to modernise and labour fought changes.

2 Overseas competition

Newly industrialising countries have the advantage of cheap labour, expanding national markets, and the newest technology. This has led to a *global shift* of manufacturing industry towards South-East Asia.

4 Rationalisation

There are a number of options facing a rationalising industry:

Decision to rationalise

Specialisation allows for economies of scale

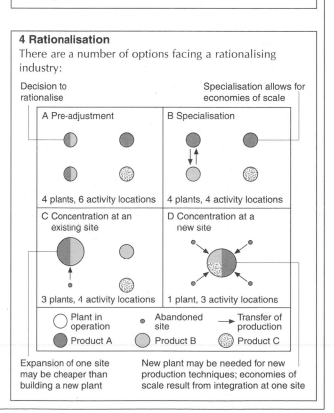

Expansion of one site may be cheaper than building a new plant

New plant may be needed for new production techniques; economies of scale result from integration at one site

Reindustrialisation

Reindustrialisation is the development of new industries which has followed deindustrialisation in many regions of the developed world. Two processes are relevant to the expansion of manufacturing industry in the developed world: the inward investment of large, multinational corporations (MNCs), and the rise of small firms.

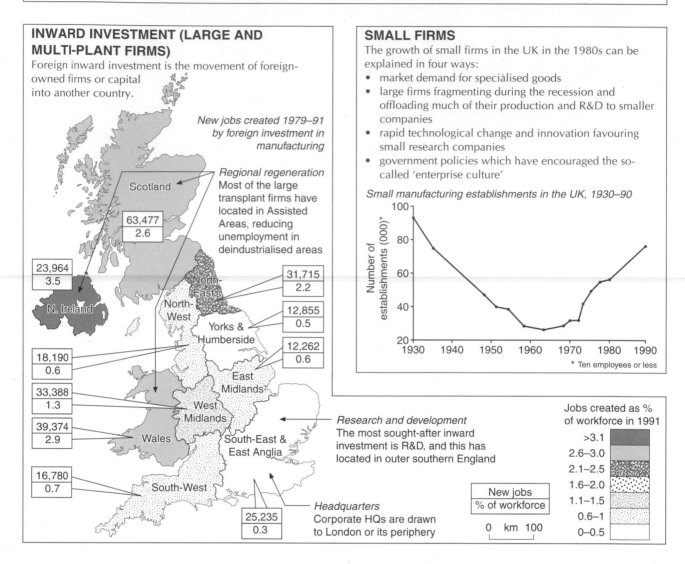

INWARD INVESTMENT (LARGE AND MULTI-PLANT FIRMS)
Foreign inward investment is the movement of foreign-owned firms or capital into another country.

New jobs created 1979–91 by foreign investment in manufacturing

Regional regeneration
Most of the large transplant firms have located in Assisted Areas, reducing unemployment in deindustrialised areas

Scotland 63,477 / 2.6

N. Ireland 23,964 / 3.5

North-East 31,715 / 2.2

North-West 12,855 / 0.5

Yorks & Humberside 12,262 / 0.6

18,190 / 0.6

East Midlands

West Midlands 33,388 / 1.3

Wales 39,374 / 2.9

South-West 16,780 / 0.7

South-East & East Anglia 25,235 / 0.3

Research and development
The most sought-after inward investment is R&D, and this has located in outer southern England

Headquarters
Corporate HQs are drawn to London or its periphery

| New jobs |
| % of workforce |

0   km   100

Jobs created as % of workforce in 1991
- >3.1
- 2.6–3.0
- 2.1–2.5
- 1.6–2.0
- 1.1–1.5
- 0.6–1
- 0–0.5

SMALL FIRMS
The growth of small firms in the UK in the 1980s can be explained in four ways:
- market demand for specialised goods
- large firms fragmenting during the recession and offloading much of their production and R&D to smaller companies
- rapid technological change and innovation favouring small research companies
- government policies which have encouraged the so-called 'enterprise culture'

Small manufacturing establishments in the UK, 1930–90

Number of establishments (000)*

(graph: x-axis 1930 to 1990, y-axis 20 to 100)

* Ten employees or less

LARGE FIRMS VERSUS SMALL FIRMS

	Large firms	Small firms
Marketing	Comprehensive distribution and servicing facilities. High degree of market power with existing products	Ability to react quickly and to keep abreast of fast-changing market requirements (Market start-up abroad can be prohibitively costly)*
Management	Experienced managers able to control complex organisations and establish corporate strategies (Can suffer an excess of bureaucracy. Managers can become cautious planners rather than entrepreneurs)*	Lack of bureaucracy. Dynamic, entrepreneurial managers react quickly to take advantage of new opportunities and are willing to accept risk
Internal communications	(Internal communication often cumbersome; this can lead to missed opportunities and poor response to market change)*	Efficient and informal communication networks. Flexible and fast response to problems. Rapid change when faced with new opportunities
Qualified technical manpower	Ability to attract highly skilled technical specialists. Can support the establishment of a large R&D laboratory	(Often lack suitable qualified technical specialists. Unable to support R&D without external help from larger company)*

* Statements in brackets represent areas of potential disadvantage

The steel industry

LOCATIONAL CHARACTERISTICS

Resources	Efficiency	Scale	Technology	Life cycle
no longer the classic Weberian industry	move towards higher productivity	modern plants are *integrated*		considerable evidence for a shift in production from the west to South-East Asia
raw materials are exhausted in many western countries	larger plants mean many older steelworks close down (rationalisation)	smelting of iron ore and conversion to steel carried out in the same plant		many western steel plants are tied to raw material locations which are now obsolete
raw materials now come from overseas (USSR, China, Brazil, Australia)	also less people are employed in each plant	large, new plants favour greenfield sites and coastal locations		developing world nations are industrialising and using the most modern technology giving them a *comparative advantage*
break-of-bulk favours coastal locations	can lead to regional problems	some arguments that integrated plants are inflexible and that mini-mills using scrap are more efficient		

THE BRITISH STEEL INDUSTRY

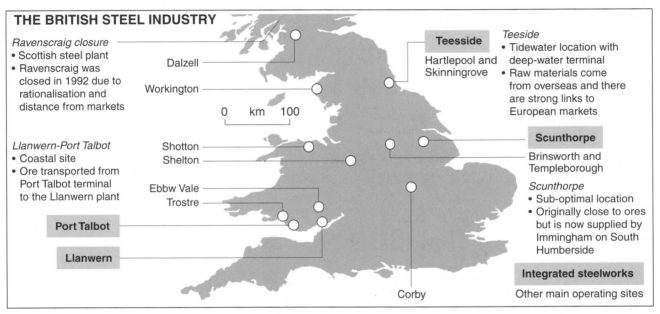

Ravenscraig closure
- Scottish steel plant
- Ravenscraig was closed in 1992 due to rationalisation and distance from markets

Llanwern-Port Talbot
- Coastal site
- Ore transported from Port Talbot terminal to the Llanwern plant

Dalzell
Workington
Shotton
Shelton
Ebbw Vale
Trostre
Port Talbot
Llanwern
Corby

Teesside
Hartlepool and Skinningrove

Teeside
- Tidewater location with deep-water terminal
- Raw materials come from overseas and there are strong links to European markets

Scunthorpe
Brinsworth and Templeborough

Scunthorpe
- Sub-optimal location
- Originally close to ores but is now supplied by Immingham on South Humberside

Integrated steelworks
Other main operating sites

0   km   100

CHANGES IN THE BRITISH STEEL INDUSTRY

Fall in demand	Competition	Privatisation	Europeanisation
recession-based cuts in the steel consuming industries	UK's production fell by over 31% between 1970 and 1991	British Steel is now a private company	entry into Europe and the Single Market
technological developments mean less steel is needed	competition from subsidised producers in the EU (Italy, Spain)	in the past steel plants were supported by state subsidies	competition has increased from overseas producers (EU and South-East Asia)
improvements in quality and durability means steel lasts longer	new producers in South-East Asia are now producing cheap, good quality steel because of low labour costs and modern plants	plants like Ravenscraig were supported for political reasons and used as part of regional policy	overcapacity has meant that the member states have been forced to cut production
substitution of steel by other materials (aluminium, plastics)		British Steel needs to please shareholders, not voters	the result has been rationalisation and restructuring with closures and lay-offs

The car industry

The world car industry can be viewed at three levels: the global, the national, and the regional. Each level is related but has different locational characteristics.

1 GLOBAL LOCATION

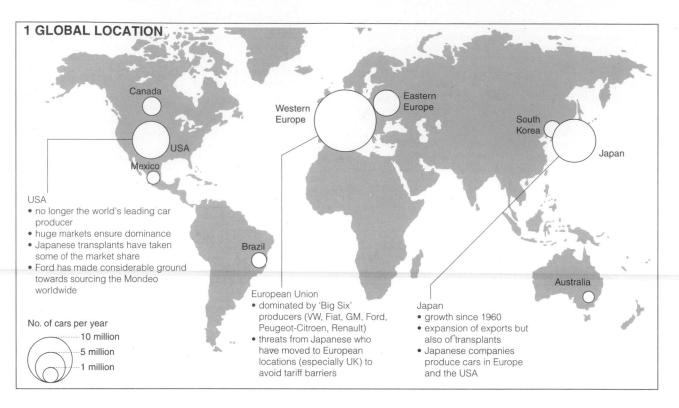

USA
- no longer the world's leading car producer
- huge markets ensure dominance
- Japanese transplants have taken some of the market share
- Ford has made considerable ground towards sourcing the Mondeo worldwide

No. of cars per year
- 10 million
- 5 million
- 1 million

European Union
- dominated by 'Big Six' producers (VW, Fiat, GM, Ford, Peugeot-Citroen, Renault)
- threats from Japanese who have moved to European locations (especially UK) to avoid tariff barriers

Japan
- growth since 1960
- expansion of exports but also of transplants
- Japanese companies produce cars in Europe and the USA

2 NATIONAL LOCATION - THE JAPANESE IN THE UK

The Japanese have shifted much of their production to Europe, especially the UK. There are two reasons for the move to Europe:
- to avoid tariff barriers and the restrictions of the Single Market

- to produce a car specifically for the European market

They have chosen the UK as their favoured location because of:
- the UK's 'open door' policy - the government and most local authorities welcome Japanese investment

- access to EU markets - if the Japanese produce cars using 60% of their components sourced from Europe they avoid tariffs

- regional assistance - support available in intermediate and development areas can be a strong incentive

- language and culture - English is the universal business language

- labour costs - the UK has amongst the cheapest labour costs in Europe

0   km   150

Nissan Sunderland

Toyota Burnaston

Honda/Rover Longbridge

Toyota Shotton

Honda Swindon

3 REGIONAL LOCATION - TOYOTA

- In 1989 Toyota invested £700 million in a car plant in Burnaston, Derbyshire.

- Toyota has 160 European-based suppliers at present, from ten countries in the EU.

- The plant operates just-in-time production and purchases parts and components worth £113 million from within 50 miles of the plant.

Greenfield location
The plant is built on a disused airfield just outside Derby

Proximity of the components firms of the East Midlands
The East and West Midlands have a long tradition of supplying components to car firms

REASONS FOR LOCATION

Skilled workforce
By the end of 1992 the plant had received more than 20,000 job applications, 50% from within a 15 mile radius

Transport links to the EU
The plant lies adjacent to the M1/M6 link to the rest of England and the EU

High-technology industry

DEFINITIONS AND CHARACTERISTICS

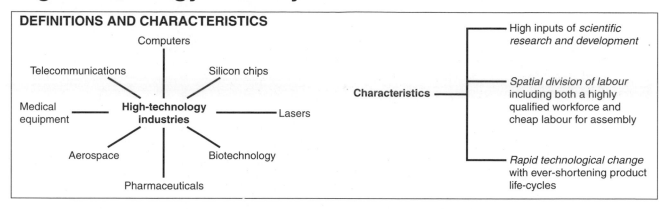

- **High-technology industries**
  - Computers
  - Telecommunications
  - Silicon chips
  - Medical equipment
  - Lasers
  - Aerospace
  - Biotechnology
  - Pharmaceuticals

Characteristics
- High inputs of *scientific research and development*
- *Spatial division of labour* including both a highly qualified workforce and cheap labour for assembly
- *Rapid technological change* with ever-shortening product life-cycles

LOCATION CHARACTERISTICS: THE BRITISH EXPERIENCE

High-technology industries have been characterised as *footloose* (not constrained in their choice of location by traditional factors, such as raw materials, energy, and transport). The major locational factor is *labour*. Skilled and well-qualified labour is needed for *research and development*. Cheap labour is also required for assembly of parts. *Transport costs* are relatively unimportant because they form a small part of the unit cost of what are small but expensive items. The British experience shows a clear *urban-rural shift*, especially in the *small*, innovative high-technology firms.

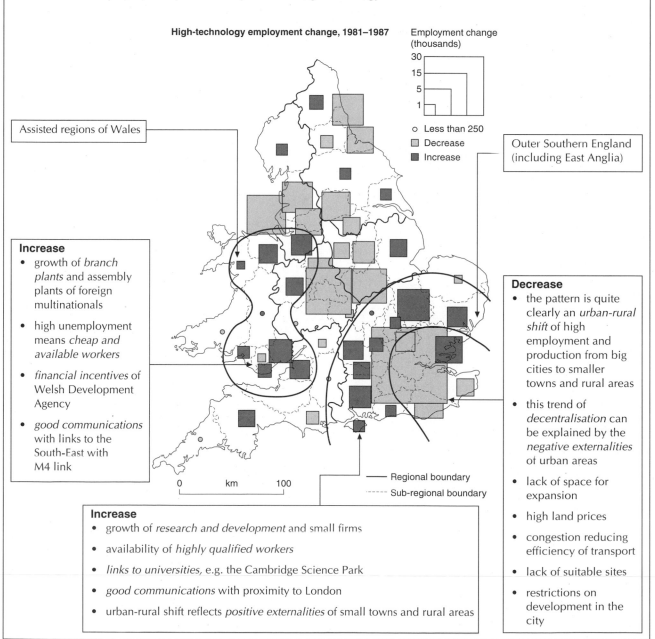

High-technology employment change, 1981–1987

Employment change (thousands)
- 30
- 15
- 5
- 1
- o Less than 250
- ☐ Decrease
- ■ Increase

Assisted regions of Wales

Outer Southern England (including East Anglia)

Increase
- growth of *branch plants* and assembly plants of foreign multinationals
- high unemployment means *cheap and available workers*
- *financial incentives* of Welsh Development Agency
- *good communications* with links to the South-East with M4 link

Decrease
- the pattern is quite clearly an *urban-rural shift* of high employment and production from big cities to smaller towns and rural areas
- this trend of *decentralisation* can be explained by the *negative externalities* of urban areas
- lack of space for expansion
- high land prices
- congestion reducing efficiency of transport
- lack of suitable sites
- restrictions on development in the city

0 km 100

— Regional boundary
--- Sub-regional boundary

Increase
- growth of *research and development* and small firms
- availability of *highly qualified workers*
- *links to universities,* e.g. the Cambridge Science Park
- *good communications* with proximity to London
- urban-rural shift reflects *positive externalities* of small towns and rural areas

Summary of locational trends in manufacturing industry

	DEINDUSTRIAL-ISATION/REINDUS-TRIALISATION	GLOBAL SHIFT	SPATIAL DIVISION OF LABOUR	MODELS	GOVERNMENT INFLUENCE	CASE STUDIES/EXAMPLES
IRON AND STEEL	declining regions due to rationalisation and competition from overseas; negative deindustrialisation	move of production from western countries to the industrialising nations of South-East Asia	large integrated steelworks mean there is little division of labour within countries; HQs do locate in London	classic Weberian industry; still relevant with location at break-of-bulk; product life-cycle explains shift to developing world	subsidies support many inefficient producers; if these are removed rationalisation and deindustrialisation result	concentration of activity at Teesside and Port Talbot-Llanwern
CAR INDUSTRY	deindustrialisation of the British car industry; reindustrialisation by Japanese car firms locating in assisted areas	Japanese producers see a three-way global expansion with bases in South-East Asia, Europe, and the USA	there is considerable division of labour in the car industry; some assembly does take place in the developing world, e.g. VW's Beetle plant in Mexico	just-in-time (JIT) production does call for close transport links; labour is important (both skilled and cheap); JIT also leads to agglomeration	protectionism in France and Italy; Italian and British regional assistance has attracted car plants to areas of regional deprivation	concentration of production in West-East Midlands but also Nissan in the North-East
HIGH-TECHNOLOGY	seen as the future of manufacturing in post-industrial societies; major element of reindustrialisation	much high-technology assembly work is undertaken in newly industrialising countries	important example of a multi-plant industry with HQs in cities, R&D in the core, and assembly in the periphery	product life-cycle explains the fast turnover of companies; constrained location theory links to the move of high-tech out of cities	assembly plants are drawn to assisted areas like South Wales and the new towns of Scotland	locational split between outer southern England and East Anglia (HQs and R&D) and assisted areas of Scotland and Wales
INWARD INVESTMENT	multinationals have been used to reindustrialise declining regions	inward investment is an important element of the movement of production from the developed to developing world	UK has been called 'Taiwan of Europe' with Japanese transplants undertaking assembly using the UK's cheap labour	transport costs are less important than an overall global strategy; sometimes this means a sub-optimal location	the UK's policy has done much to attract inward investment: • regional assistance • open door policy • development agencies	the UK is the centre of Japanese and US inward investment in the EU (over 40%)

Energy

Renewable forms of energy are not depleted by use. This includes all forms of energy generated from **non-finite sources** such as water, wind, solar, and geothermal sources. Energy can be in the form of heat or electricity and is often on a small scale, e.g. a wind pump.

Non-renewable forms of energy are depleted by use. This includes oil, gas, coal, and wood, where the rate of use exceeds the rate of formation.

The **energy mix** of a country, i.e. the amount and type of each form of energy used, depends upon:
- availability and reliability of supply
- suitability and efficiency of supply
- costs in terms of production, distribution, and use
- type of market
- political factors

WORLD ENERGY DEMAND, 1976-94

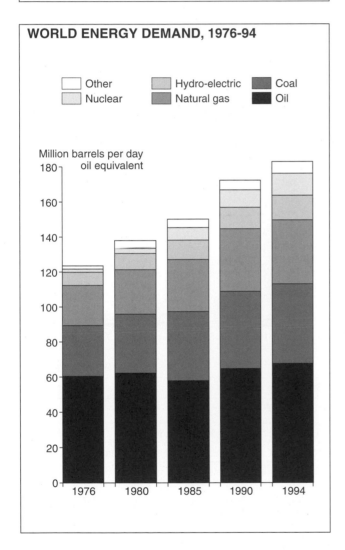

WORLD ENERGY PRODUCTION, 1990

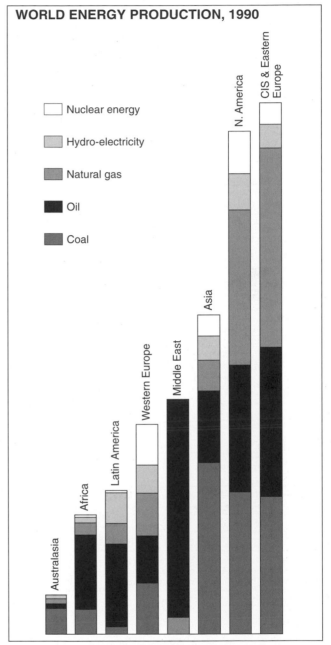

There is a high correlation between energy consumption and standard of living. However, there are many important factors to consider:
- types of economy - urban or rural, industrial or agricultural
- whether the country is a net importer or exporter of energy
- stage of economic development - e.g. NICs consume a great amount of energy
- efficiency and quality of fuel
- climatic variation - warm areas need less heating
- cost and availability of energy
- substitution of other products, e.g. alcohol for petrol

Low levels of energy consumption in LDCs are explained by:
- lack of suitable resources
- lack of economic development to finance either the rapid exploitation of energy resources or imports of energy
- rapid growth of population (demand > supply)
- lack of capital to develop alternative forms of energy
- debt
- lack of technological resources
- lack of trust, especially with regard to nuclear power
- lack of fuelwood

Coal-mining in the UK

The factors affecting coal-mining include **demand, competition, grade of coal, geological conditions, availability of capital, technology**, and, increasingly, **environmental impact**.

Britain's coal-mining industry has been severely rationalised since 1945. The demand for coal has dropped, as has the number of operational collieries and coal-miners. A number of reasons explain this:

- falling demand and the changing nature of demand

- improvements in mechanisation and technology

- the government's decision to buy cheap Middle East oil

- the development of North Sea oil and gas

- the attempt by the National Coal Board in 1974 to modernise the coal-mining industry

- cheaper coal in the 1980s from South Africa, the USA, and Australia

- the miners' strike of 1985 which lost them government support

- increasing environmental concerns concerning sulphur dioxide, carbon dioxide, acid rain, air and water pollution, and subsidence

During the 1990s decline accelerated. The government announced further closures and redundancies. By the end of 1994, when British Coal was privatised, it had become much more competitive and streamlined. Peripheral coalfields were closed down, leaving higher quality, more efficient central ones.

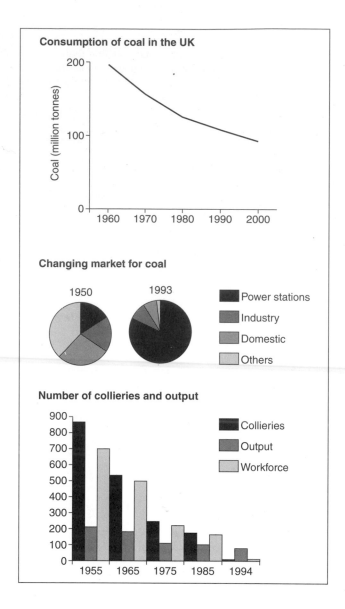

Consumption of coal in the UK

Changing market for coal

Number of collieries and output

A comparison of two British coalfields: Asfordby, a thriving colliery, and Sharlston, a peripheral one		
	CENTRAL	PERIPHERAL
Example	Asfordby, East Midlands	Sharlston, near Wakefield, West Yorkshire
Geology	Little folding or faulting, thick seams	Contorted strata, difficult to work, thin seams
Reserves	Large supplies	Mostly exhausted
Quality	Good	High
Supplies	Plentiful	Largely exhausted
Costs	Low	High
Mechanisation	Easy	Difficult
Productivity	High	Low
Market	Electricity generation	Exports, 'traditional' heavy industries
Overseas competition	Limited	Considerable
Regional economies	Diversified	Dependency on coal and related traditional industries
Unemployment	Below UK average	Well above average
Environment	Strict controls, evidence less obvious	Highly polluted: tips, slag heaps
Trends	Decline halted, future investments concentrated on these areas	Rapid and serious decline, major future investments unlikely
Prospects	Good: concentration on best reserves	Bleak: much smaller coal industry

Oil production

PRODUCTION AND CONSUMPTION

The OPEC countries (Organisation of Petroleum Exporting Countries) of the Middle East account for a large part of the world's **oil production**. The Old Industrialised Countries (OICs) provide about 23% of world production. By contrast, **oil consumption** is mostly in the OICs (48%) and Asia and Australasia (20%). Consumption largely reflects levels of economic development.

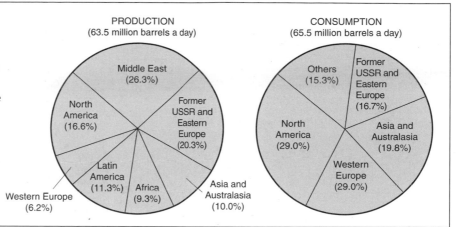

PRODUCTION
(63.5 million barrels a day)

Middle East (26.3%)
Former USSR and Eastern Europe (20.3%)
Asia and Australasia (10.0%)
Africa (9.3%)
Latin America (11.3%)
Western Europe (6.2%)
North America (16.6%)

CONSUMPTION
(65.5 million barrels a day)

Former USSR and Eastern Europe (16.7%)
Asia and Australasia (19.8%)
Western Europe (29.0%)
North America (29.0%)
Others (15.3%)

OIL REFINING

About 80% of oil refining takes place in the **triad** of North-West Europe, North America, and Japan. Refineries are no longer associated with the oil fields as in the early days of development. Today they are increasingly found close to markets and in coastal locations. **Relocation** is due to a number of factors:

- change in size, location, and nature of markets, i.e. markets are increasingly larger and include a greater variety of manufactured products (such as synthetic fibres and pharmaceuticals - not just oil and petrol)
- improvements in transport, e.g. pipelines and VLCCs (very large crude carriers)
- costs of building and maintaining a refinery are so great that it is sensible to locate close to the market
- political interference in the form of tariffs

Market-based refineries safeguard fuel supplies, save foreign currency, and provide an export industry in manufactured goods.

Oil reserves are generally found in geological structures such as anticlines, fault traps, and salt domes. At present rates of production and consumption, known reserves could last for another forty years. Nearly two-thirds of the world's reserves are found in the Middle East.

Percentage of world oil reserves	
Middle East	62.5
Latin America	12.5
Former USSR and Eastern Europe	5.9
Africa	5.9
Asia and Australasia	4.5
North America	4.2
Western Europe	1.8

EXPLORATION AND DEVELOPMENT OF RESOURCES

Factors affecting the exploration and development of resources since 1973 include:
- the 1973 oil crisis and the subsequent rapid increase in oil prices
- the continued growth in energy demands resulting from continued population growth, economic development, and rising living standards
- the increased availability of technology, leading to searches in previously difficult terrain
- government backing and EU funding, which has stimulated new developments

However, attempts to develop energy resources are often constrained because:
- national governments may take royalties, reducing profits and limiting the desire for future projects
- there's an increasing need for energy conservation
- increasing energy efficiency has reduced demand
- the world recession has reduced oil prices and demand
- OPEC still controls pricing and the rate of development of reserves

MIDDLE EAST OIL: GEOGRAPHIC IMPLICATIONS

The importance of the Middle East as a supplier of oil is critical - OPEC controls the price of crude oil, and this has increased its economic and political power. It has also increased dependency upon the Middle East by all other regions. This provides an incentive for OICs to increase energy conservation or develop alternate forms of energy.

Countries therefore need to:
- maintain good political links with the Middle East
- involve the Middle East in economic co-operation

It also means that there is a need for political stability in the Middle East and a need to reassess coal and nuclear power as energy options.

ENVIRONMENTAL IMPLICATIONS

- Oil slicks from tankers, e.g. the Torrey Canyon (1967), Exxon Valdez (1989), Braer (1993), and Sea Empress (1996).
- Damage to coastlines, fish stocks, and communities dependent upon the sea.
- Water pollution caused by tankers illegally washing/cleaning out in the North Sea.
- Gulf War damage - oil wells and storage sites can be targets for destruction, resulting in immeasurable environmental damage.

Renewable energy

The main types of renewable energy are HEP and tidal, wind, solar, and nuclear power. Others include geothermal power and the use of biogas.

HYDROELECTRIC POWER (HEP)

HEP is a renewable form of energy that harnesses fast-flowing water with a sufficient head.

The **location** of HEP stations depends upon:
- relief
- geology
- river regime
- climate
- market demand
- transport facilities

The **site** depends upon:
- local valley shape (narrow and deep)
- local geology (strong, impermeable rocks)
- lake potential (a large head of water)
- local land use (non-residential)
- local planning (lack of restrictions)

However:
- HEP plants are very costly to build
- only a small number of places have a sufficient head of water
- markets are critical - HEP stations are often associated with aluminium smelters to use up excess energy

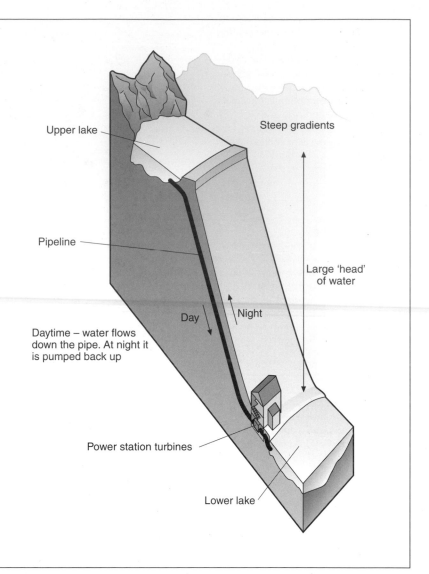

Steep gradients

Upper lake

Pipeline

Large 'head' of water

Day   Night

Daytime – water flows down the pipe. At night it is pumped back up

Power station turbines

Lower lake

NUCLEAR POWER

Advantages	Constraints
• A cheap, reliable, and abundant source of electricity	• Recession in the late 1980s and early 1990s has reduced the demand for energy: less energy development is now required
	• Oil prices have fallen and the coal industry is more competitive; the cost of nuclear energy has risen due to more stringent safety measures
• Unlike coal and oil, which have reserves estimated to last 300 years and 40 years respectively, there is a plentiful supply of uranium, enough for it to be considered a renewable form of energy	• The EU has a diverse range of energy suppliers: the threat of disruption to any one source is therefore less worrying than it used to be
	Disadvantages
• Uranium fuel is available from countries such as the USA, Canada, South Africa, and Australia, and so Western Europe would not have to rely on unstable political regions, such as the Middle East, for its energy needs	• Uranium is a radioactive material and so the nuclear power industry is faced with the hazards of waste disposal and the problems of decommissioning old plants and reactors
• The European Union is in favour of nuclear power and estimates that 40% of the EU's electricity will be provided by nuclear power in 2000 (15% of total energy)	• Rising environmental fears concerning the safety of nuclear power and nuclear testing are based on experience: disasters happen, as at Chernobyl in April 1986

TIDAL POWER

Tidal power is a renewable, clean energy source that requires a funnel-shaped estuary with a large tidal range. It also needs to be free of other developments. One of the best-known tidal power schemes is on the Rance Estuary in Brittany. The proposed Severn tidal barrage is unlikely to be built owing to its cost. Large-scale production is limited for a number of reasons:

- high cost of development (cost of proposed Severn barrage exceeds that of the Channel Tunnel)
- limited number of suitable sites
- environmental damage to estuarine sites
- long period of development
- possible effects on ports and industries upstream

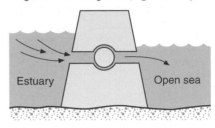

RISING TIDE – sluice gates in barrage opened to allow water to flood into the estuary

Estuary   Open sea

FALLING TIDE – water flows out of the estuary, turning turbines and generating electricity

Estuary   Open sea

WIND POWER

Orkney & Shetland Isles

4

1,2,3

▣ Windiest parts of Britain

Wind turbines
1 ⎫
2 ⎬ Burgar Hill, Orkney
3 ⎭
    ⎧ 3,000 kW
    ⎨ 250 kW
    ⎩ 300 kW
4 Susetter Hill, Shetland 750 kW
5 ⎫
6 ⎬ Carmarthen Bay, S. Wales
7 ⎭
    ⎧ 300 kW
    ⎨ 300 kW
    ⎩ 130 kW
8 Ilfracombe, Devon 250 kW
9 Treculliacks, Cornwall 150 kW
10 Richborough, Kent 1,000 kW
11 Offshore, Norfolk 750 kW

Possible wind farm locations
A Capel Cynon, Dyfed, Wales
B Cold Northcott, Cornwall
C Langdon Common, County Durham
D Eaglesham, Strathclyde

0   km   150

Wind power is good for small-scale production. It needs an exposed site, such as a hillside, flat land, or coastal position, and strong, reliable winds, e.g. Carmarthen Bay, Wales, and Altamont Pass, California.

Advantages:
- no pollution of air, ground, or water
- no finite resources involved
- reduces environmental damage elsewhere

Disadvantages:
- visual impact
- noisy
- winds may be unreliable

Large-scale development is hampered by the high cost of development, the large number of wind pumps needed, and the high cost of new transmission grids as suitable locations for wind farms are normally quite distant from centres of demand.

SOLAR POWER

Solar power schemes have been developed in California and Israel.

Advantages:
1 no finite resources involved - less environmental damage
2 no atmospheric pollution
3 suitable for small-scale production

Disadvantages:
1 affected by cloud, seasons, darkness
2 not always possible when demand exists

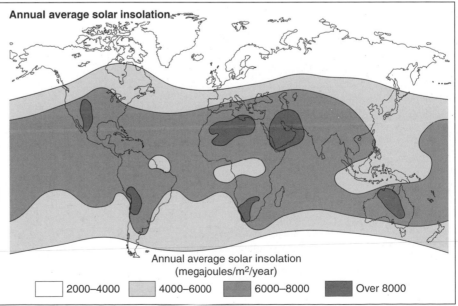

Annual average solar insolation

Annual average solar insolation (megajoules/m^2/year)

▢ 2000–4000   ▨ 4000–6000   ▩ 6000–8000   ■ Over 8000

Transport

Factors affecting type of transport used include:
- item to be transported
- cost of transporting it
- speed with which it needs to be transported, e.g. perishable goods such as flowers and fruit need to be transported rapidly whereas bulky goods such as coal can be transported by the cheapest means possible

Economies of scale are also important: it is cheaper to carry bulk than small amounts and therefore bulk carriers are increasingly used.

A COMPARISON OF FORMS OF TRANSPORT

	Advantages	Disadvantages
Road	<ul><li>fast over short routes/motorways</li><li>flexible routes/well developed network</li><li>door to door</li><li>independence and privacy</li></ul>	<ul><li>expensive over long distances</li><li>slow in urban areas</li><li>costs increase rapidly with time</li><li>expensive to build and maintain</li><li>air pollution</li><li>structural effect</li><li>limited load size/only small loads can be carried</li></ul>
Rail	<ul><li>fast over medium-to-long journeys</li><li>cost effective over medium-to-long journeys</li><li>can carry hundreds of passengers and heavy bulky goods</li><li>safe, dependable, and comfortable</li></ul>	<ul><li>expensive on short and long routes</li><li>limited to routes</li><li>limited by physical geography (gradient)</li><li>expensive to build and maintain</li></ul>
Water (ocean)	<ul><li>economical, cheap over long distance</li><li>no cost in building the route</li><li>good for bulky, low cost goods, e.g. coals, ores, grains</li></ul>	<ul><li>slow</li><li>very limited routes to deep-water ports</li><li>ships expensive to build and maintain</li><li>environmental problems - especially pollution</li><li>ports take up a great deal of space</li></ul>
Inland canals	<ul><li>cheap over long distances</li><li>reasonable for bulky goods and good for recreation, but little else</li></ul>	<ul><li>few routes</li><li>narrow</li><li>expensive to build and maintain</li></ul>
Air	<ul><li>fast over long distances</li><li>limited congestion</li><li>good for high value transport</li><li>good for people, high-tech industries, and urgent cargo</li></ul>	<ul><li>lots of land needed for airports</li><li>noise and visual pollution</li><li>very expensive to build and maintain</li><li>no flexibility of routes</li></ul>
Pipelines	<ul><li>continuous flow</li><li>fast</li><li>cheap maintenance</li><li>no hold ups or congestion</li><li>useful for the bulk liquids, e.g. oil and gas</li></ul>	<ul><li>very expensive to build</li><li>limited to very level routes</li><li>environmental problems, e.g. the Trans-Alaska pipeline</li><li>inflexible once laid</li></ul>

Transport: topological maps

Nodes or vertices (v) are points on the map. Links or edges (e) are the lines or routes between them.

The **Shimbel index** ($\sum e$) is the total number of edges needed to connect a place with all other places. A low Shimbel index indicates high **centrality** or **accessibility**.

The **beta index** (e/v) is a measure of connectivity:
- the higher the value of e/v the more connected the network
- in simple networks e/v < 1.0
- a network with a complete circuit has a value of 1.0
- an integrated, developed network has a value of > 1.0

The **cyclomatic number** (c) measures the number of complete networks in a system.
c = (e - v) + 1
A high value cyclomatic number indicates more complete circuits.

The **alpha index** compares the actual number of circuits with the maximum possible for that network.

$$\text{Alpha index} = \frac{(e - v + 1)}{2v + 5} \times 100\%$$

The result is expressed as a percentage. A high alpha index indicates a high degree of **connectivity** whereas a low value indicates very limited connectivity. A value of 100% indicates complete connectivity.

TYPES OF NETWORKS

No links

Connected - but not necessarily to each other

Circuit: links all points

Tree: all points linked but no circuit

Complete: all points linked to each other

Sub-graph: part of system is detached

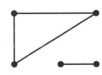

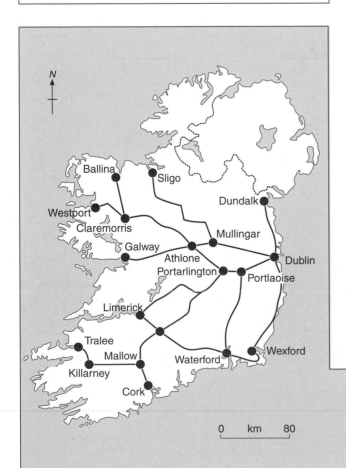

N

0   km   80

AN EXAMPLE OF A TOPOLOGICAL NETWORK: THE IRISH RAILWAY SYSTEM

- lowest Shimbel index — Portarlington 42
- highest Shimbel index — Tralee 92
- Beta index — 20/19
- Cyclomatic number — (20 - 19) + 1 = 2
- Alpha index — 4.65%

Transport policies

THE M3 EXTENSION AT TWYFORD DOWN

The extension is 6 km long. Much of the area had been designated an Area of Outstanding Natural Beauty (AONB) and there were also two Sites of Special Scientific Interest (SSSIs) and two ancient monuments. The 6 km route cost £35 million to build.

For 'This road is vital for the prosperity of the South of England.'
- Winchester bypass is no longer used - it was a source of congestion.
- The M3 is now more efficient - 5-10 minutes have been taken off the journey time to London.
- Verges have been seeded for environmental reasons.

Against
- Each kilometre of road requires more than 150,000 tonnes of rock.
- More traffic leads to increased greenhouse gases.
- 125 m bare chalk cutting.
- 6 ha of land are used for every kilometre of motorway.
- Pollutants from the road, such as oil, petrol, and salt, may contaminate nearby water meadows.
- Iron Age settlements, rare orchids and butterflies, and Downland ecology disrupted.

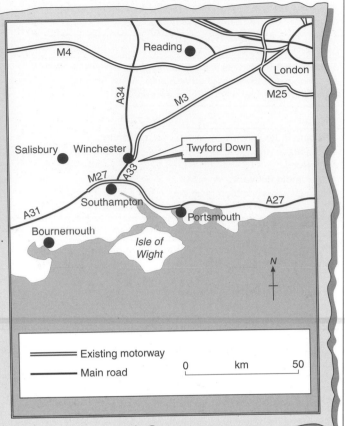

OXFORD'S PARK AND RIDE SCHEME

Oxford's parking problem is severe. Up to 80,000 vehicles a day converge on the city and its 15,000 parking spaces. At peak times, the three-mile trip from the Pear Tree roundabout to the city centre is 30 minutes of congestion, delay, frustration, and, increasingly, air pollution. Asthma has been linked to air pollution. And Oxford has one of the highest rates of asthma in the UK. During the summer of 1995, levels of carbon monoxide and nitrogen exceeded WHO safe limits. Even in nearby rural areas photochemical ozone reduced air quality.

In 1973, Oxford City Council opened the Redbridge Park and Ride scheme, accommodating 200 cars. By 1995 four sites provided 3750 parking spaces and plans were published to provide an extra 1000 spaces. Congestion and pollution had forced the planners to develop the Park and RIde scheme.

- Over 80,000 vehicles a day enter Oxford.
- Over 35% of visitors to central Oxford arrive by car.
- Park and Ride intercepts 18% of Oxford-bound cars during off-peak times and 38% at peak times.
- There are over 15,000 parking spaces in Oxford. Park and Ride has added extra parking spaces rather than replaced city centre spaces.
- Oxford has the highest bus usage for a city its size and the second highest use of bicycles.
- Park and Ride is having an adverse effect on rural bus services as over 1000 potential bus users use Park and Ride instead.
- Park and Ride has eased the pressure on city centre roads and car parks.
- Park and Ride car parks take up 12 ha of good quality green belt land.

There needs to be:
- restricted parking in the city centre
- more bus routes and bus lanes
- payment by users of Park and Ride so that revenue is raised to pay for public transport. A fee of 50p per day would generate £500,000 each year

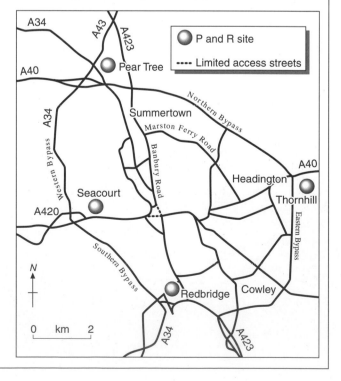

Regional inequalities

TYPES OF REGIONAL PROBLEM

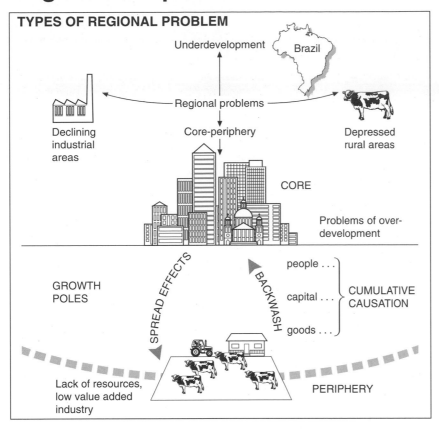

Underdevelopment — Brazil

Regional problems

Core-periphery

Declining industrial areas

Depressed rural areas

CORE

Problems of over-development

GROWTH POLES

SPREAD EFFECTS

BACKWASH

people . . .
capital . . .
goods . . .

CUMULATIVE CAUSATION

Lack of resources, low value added industry

PERIPHERY

MEASURING REGIONAL INEQUALITIES

A variety of economic, social, demographic, and environmental indices are used to measure regional variations in development and standard of living: Gross Domestic Product, unemployment rates, labour costs, income per head, proportion of population employed in agriculture, type of industry, receipt of regional and structural aid, inward investment, rates of out-migration, levels of education, housing quality, social service provision, and environmental dereliction.

EXAMPLES OF REGIONAL DISPARITIES

Local: disparities between the inner city areas and rich suburbs.

National: Britain's North-South Divide, Italy's Mezzogiorno.

International: core and periphery in the European Union, e.g. Portugal's position in the EU.

Global: the developed west versus the less developed countries.

FINGLETON'S POSITIVE AND NEGATIVE FEEDBACK (1991)

There are two views of regional inequality and regional policy:

Negative feedback (neo-classical) suggests market forces are all that are needed to draw investment to peripheral areas and attract labour from core areas.

Positive feedback (cumulative causation) suggests continual divergence due to a region's comparative advantage which continues to attract investment and labour.

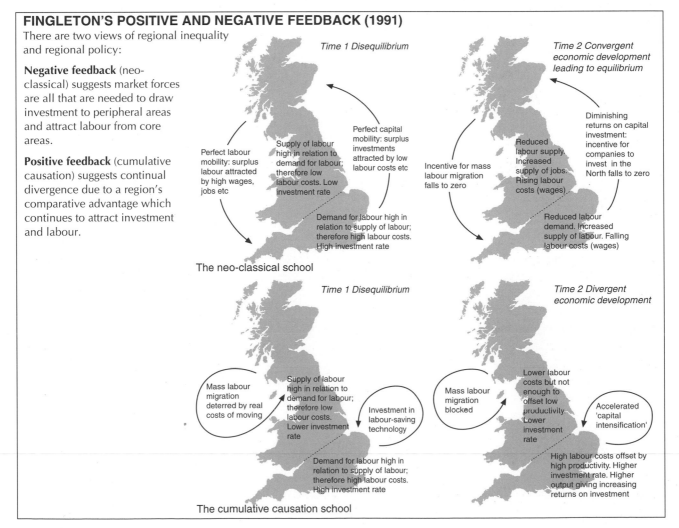

Time 1 Disequilibrium

Perfect labour mobility: surplus labour attracted by high wages, jobs etc

Supply of labour high in relation to demand for labour; therefore low labour costs. Low investment rate

Perfect capital mobility: surplus investments attracted by low labour costs etc

Demand for labour high in relation to supply of labour; therefore high labour costs. High investment rate

The neo-classical school

Time 2 Convergent economic development leading to equilibrium

Incentive for mass labour migration falls to zero

Reduced labour supply. Increased supply of jobs. Rising labour costs (wages).

Diminishing returns on capital investment: incentive for companies to invest in the North falls to zero

Reduced labour demand. Increased supply of labour. Falling labour costs (wages)

Time 1 Disequilibrium

Mass labour migration deterred by real costs of moving

Supply of labour high in relation to demand for labour; therefore low labour costs. Lower investment rate

Investment in labour-saving technology

Demand for labour high in relation to supply of labour; therefore high labour costs. High investment rate

The cumulative causation school

Time 2 Divergent economic development

Mass labour migration blocked

Lower labour costs but not enough to offset low productivity. Lower investment rate

Accelerated 'capital intensification'

High labour costs offset by high productivity. Higher investment rate. Higher output giving increasing returns on investment

Regional disparities in Italy

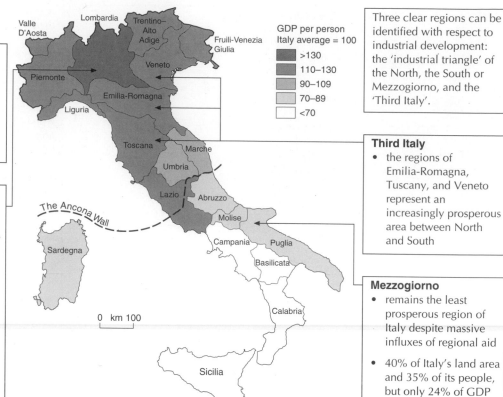

Three clear regions can be identified with respect to industrial development: the 'industrial triangle' of the North, the South or Mezzogiorno, and the 'Third Italy'.

In terms of national growth, Italy has recently overtaken Britain as the fifth-biggest industrial power of the west. However, this expansion has not benefited the whole country.

GDP per person
Italy average = 100
- >130
- 110–130
- 90–109
- 70–89
- <70

Industrial triangle
- the cities of Milan, Turin, and Genoa enclose one of the EU's most affluent regions
- northern Italy has been compared with Japan and Taiwan in its rapid and sophisticated industrial expansion
- products include textiles, machine tools, leather goods, footwear, and fashion clothes

Third Italy
- the regions of Emilia-Romagna, Tuscany, and Veneto represent an increasingly prosperous area between North and South

Mezzogiorno
- remains the least prosperous region of Italy despite massive influxes of regional aid
- 40% of Italy's land area and 35% of its people, but only 24% of GDP

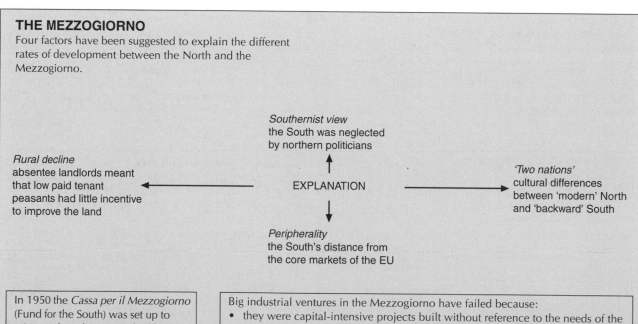

THE MEZZOGIORNO

Four factors have been suggested to explain the different rates of development between the North and the Mezzogiorno.

Southernist view
the South was neglected by northern politicians

Rural decline
absentee landlords meant that low paid tenant peasants had little incentive to improve the land

← EXPLANATION →

'Two nations'
cultural differences between 'modern' North and 'backward' South

Peripherality
the South's distance from the core markets of the EU

In 1950 the *Cassa per il Mezzogiorno* (Fund for the South) was set up to raise southern living standards. Two main policies were pursued:

1. land reform, which broke up the large semi-feudal estates of the South creating about 120,000 new, small farms

2. growth pole strategy, which legislated that 60% of all state investment went to the South with investments in large steel and chemical plants ('cathedrals in the desert')

Big industrial ventures in the Mezzogiorno have failed because:
- they were capital-intensive projects built without reference to the needs of the local economy or to the markets for steel and petrochemicals, which were in the North or outside Italy
- the OPEC price-hike, which quadrupled the price of oil in 1973, raised the price of raw materials for the new industries
- there is overcapacity in the European steel market

There were some successes, but growth pole strategy was abandoned in 1986.

New policy is to attract private investment through grants and subsidies. Fiat's plant in Melfi has resulted from this policy.

Reindustrialising the Ruhr

The Ruhr is the classic example of an old industrial region in transition. It was once the foremost centre of coal and iron and steel production in Europe.

- Between 1850 and 1914 the Ruhr's workforce increased from about 12,000 to 400,000.

- Deindustrialisation and restructuring between 1984 and 1994 saw a 20% fall in the Ruhr's workforce.

- Industrial decline was due to overseas competition, overcapacity, and technological change.

- The results have been unemployment, out-migration, and the need for regional growth policies.

Investment in new autobahns, airports, telecommunications, and waterfronts.

New towns and industrial complexes.

Coalfields
Many pit closures and job losses. Deeper mines with increasing use of robotics and technology, and higher productivity.

Petrochemicals and textiles
Growth in oil refining and subsequent agglomeration of petrochemicals, textiles, and pharmaceuticals.

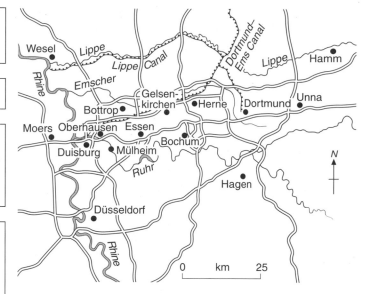

Industrial diversification
Decline in traditional mining, steel, and heavy engineering together with new developments in light, high-tech industries and services have resulted in a wider industrial base in the Ruhr.

Steel
Production down. Rationalisation and investment common.

Despite diversification of industries, the Ruhr has 13.1% unemployment.

POLICIES

1 Funds from the EU
- the Ruhr is designated an Objective Two region (economic decline and high unemployment)

- the Ruhr received over £270 million between 1989 and 1991

2 Funds from government
- the considerable funding from the German government over the 1970s and 1980s is increasingly stretched due to the burden of developing the old east Germany

3 Locally-based initiatives
- the regional government of North Rhine-Westphalia launched an initiative to promote innovation, new technologies, and new training

- the regional government also set up the Emscher Park Planning Company in 1988 which is a 10-year programme covering 17 towns and two million people - the aim is to re-landscape the most depressed part of the region and attract high-tech industries

4 Private sector initiatives
- 35 businessmen joined forces to found the Initiativkreis Ruhrgebiet

- each member contributed DM 1.5 million to form a lobbying group

RESULTS OF POLICY

The overall policy has been to slowly restructure and rationalise the iron and steel and coal industries, closing plants while at the same time providing alternative employment.

1 Environmental industries
The Ruhr's first-hand experiences of environmental problems have given rise to a successful specialisation in environmental technology.

- more than 150,000 people now work in environment-related industries

- between 1990 and 1994 DM 6 million was invested in 4500 environmental companies in the region

2 Technology transfer offices (TTOs)
- TTOs are designed to provide marketing advice and spread new ideas. TTOs are strongly linked to universities and research institutions

- TTOs have stimulated the establishment of new industries by breaking through the bias towards coal and iron and steel related growth

Portugal and the European Union

PORTUGAL IN 1986

Portugal joined the European Union in 1986. At this time it was *peripheral* both spatially and economically. The question is whether Portugal's membership of the EU and its subsequent grants from the EU's structural funds have benefited the country.

Portugal's weaknesses were identified by the European Commission in 1986:

- an inadequate economic infrastructure with regards to communications, telecommunications, and energy
- a poorly skilled or unskilled workforce
- regional imbalances

The country received 17.5% of the EU's structural funds between 1988 and 1995.

REGIONAL POLICY

As a country, Portugal is peripheral in the EU. Within the country, there are regional imbalances:

- coastal-interior - coastal areas are more urbanised and more industrialised; the interior is the opposite
- north-south - the north is more religious, politically more conservative, and more rural

Restructuring

- EU support only goes to the most successful companies
- the Vale do Ave textile region will lose 120 firms and more than 30,000 jobs
- EU funds will be directed to retraining schemes and attracting new industry

Inward investment

- Setubal has been transformed by foreign investment from a depressed shipbuilding area; investors include Ford Electronica, GM, Delco Remi, Ford, VW, and Auto Europa

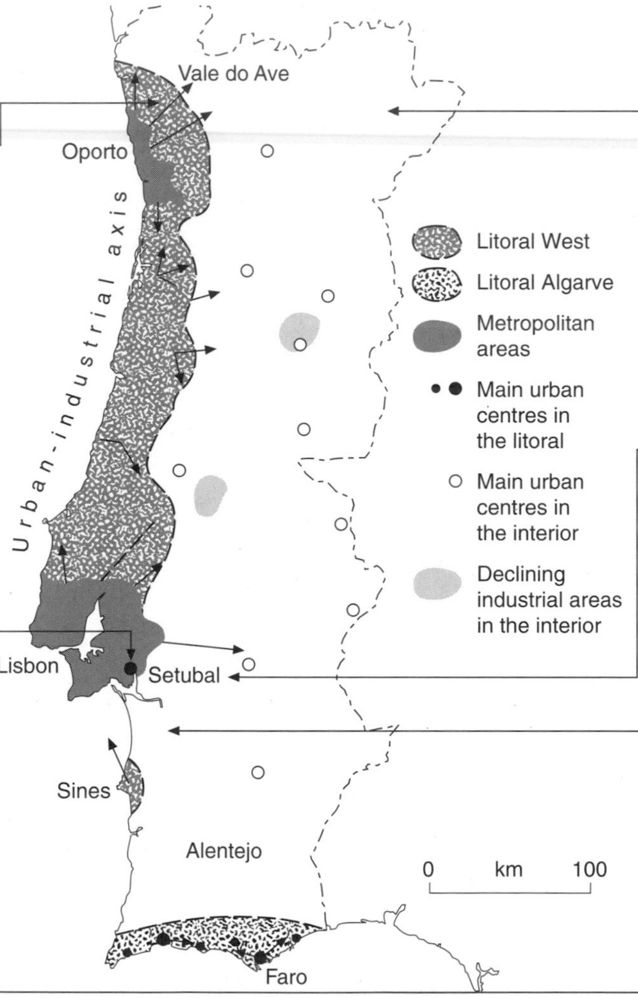

Vale do Ave
Oporto
Urban-industrial axis
Lisbon
Setubal
Sines
Alentejo
Faro
Litoral West
Litoral Algarve
Metropolitan areas
Main urban centres in the litoral
Main urban centres in the interior
Declining industrial areas in the interior
0 km 100

Northern agriculture

- small family farms have not benefited from structural funds
- since 1993 competition from other farmers in the EU has led to decline and out-migration

Southern agriculture

- larger cereal farms have won EU aid
- areas around Lisbon also benefit from proximity to the large urban market

Infrastructure

- improved links between Oporto and Lisbon and between Portugal and Spain
- less money spent on linking the coast with the interior

EVALUATION OF REGIONAL POLICY

Portugal's situation has improved since 1986 - with some of the highest growth rates and lowest levels of unemployment in Europe, it is an emerging European state with a GDP that has now risen above that of Greece. Structural funds have brought jobs and investment to Portugal. The funds have also led to a general modernisation of the country's industry and infrastructure, especially in the coastal urban-industrial areas.

The effects have been less positive on the regional imbalances within the country. The inequality between Alentejo and the richer coastal strip joining Lisbon and Oporto has grown wider.

- Alentejo is an agricultural region for which the Agricultural Guidance Fund has done little.
- Increased competition from other member states has led to a decline in agriculture.
- Most infrastructural development and inward investment is restricted to the coastal strip, especially around Lisbon.

Regional policy in the UK

There has been a long history of regional policy in the UK. The policies put forward have attempted to encourage the growth of manufacturing in the deindustrialising regions of the North. The map of regional assistance has changed between 1970 and the present. The changes reflect a shift in political ideology in addition to the changing fortunes of the UK's regions.

The three maps illustrate not only a change in where and how much aid is given but also how it is used. In the 1970s much aid was used to support British industries. Today assisted areas try to attract inward investment from foreign multinationals.

THE CHANGING MAP OF REGIONAL ASSISTANCE, 1979–1995

Policy: 1979
- large areas were given regional assistance (the North, Wales, Scotland, Northern Ireland)
- reflects an ideology that regional disparities will increase without financial aid (carrots) and disincentives (sticks)

Policy: 1984
1981 - policy of disincentives for location in the South abolished
1984 - reduction in the geographical scale of aid
1985 - reduction in the financial scale of aid (£584 million in 1985 to £384 million in 1992)

Policy: 1993
- further reduction of the regional extent of aid
- shift in regional policy focus towards the South
- more than half of the 32 new assisted areas are in the South, including areas of London (Lea Valley and Park Royal)

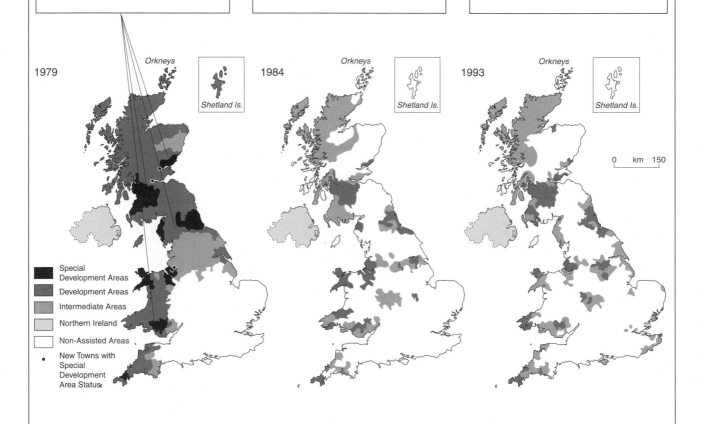

Special Development Areas
Development Areas
Intermediate Areas
Northern Ireland
Non-Assisted Areas
• New Towns with Special Development Area Status

Deindustrialisation
- northern conurbations were tied to industrial regions based on coal and mature industries
- these were declining from the 1930s onwards but were particularly hit by the recession of the 1970s

Change in ideology
- reduction in scale and amount of aid was the result of a shift in political ideology
- regional aid still important, but now controlled by competing regional development agencies for Scotland, Wales, and Northern Ireland

Recession-hit South
- southern economy based on services was hit by the recession of the late 1980s
- policies which had encouraged industries to locate outside the South-East meant that the decline in services resulted in a growth in southern unemployment

CASE STUDY: REGIONAL POLICY IN THE WEST MIDLANDS

The West Midlands can be viewed as a peripheral region with its low skills base and inner city areas containing some of the most serious problems in the country. However, it can also be viewed as the manufacturing core of Britain; an area which has attracted substantial inward investment and with competitive and low cost industries as a result of recent rationalisation. It is also unclear whether the region belongs to the North or South of the UK. Traditionally, the West Midlands is the first region to enter a recession but also the first to emerge from one.

The UK government

- Birmingham and Wolverhampton have *Development Area Status* and have their own *Urban Development Corporations* which use government money to encourage private sector investment
- an East-West corridor from Rugby through to Coventry, Birmingham, and the Black Country is classed as an *Intermediate Assisted Area*

The European Commission

- in 1988 large areas of the West Midlands were granted *Objective 2 Status* (regeneration) which triggers assistance from the *European Regional Development Fund* (ERDF)
- other areas receive funds under *Objective 3* (long-term unemployment), *Objective 4* (youth training), and *Objective 5b* (regeneration of rural areas)

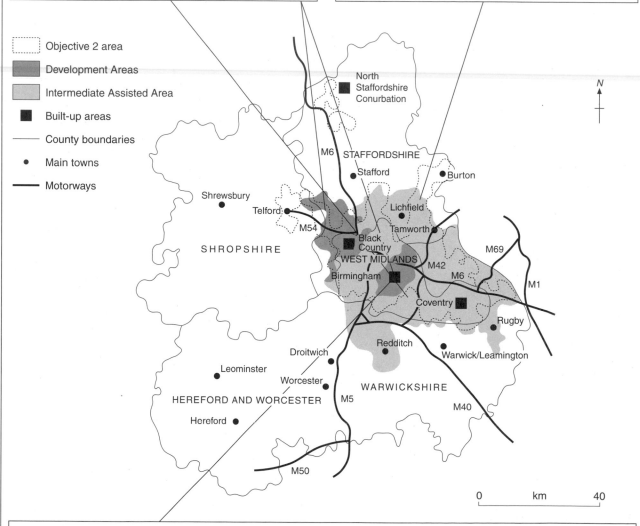

Key:
- Objective 2 area
- Development Areas
- Intermediate Assisted Area
- ■ Built-up areas
- County boundaries
- • Main towns
- Motorways

0 km 40

Investing in Jaguar

Jaguar cars is the largest single recipient of aid in the UK. The two plants - Castle Bromwich in Birmingham and Browns Lane in Coventry - shared £9.4 million in 1994. In 1995 a huge government grant of £80 million persuaded Jaguar to upgrade its plants in the West Midlands.

This will create 1300 jobs by 1999. The aid package shows the continued belief that large manufacturing plants can have a multiplier effect in stimulating regional growth.

The aid package:
1. £48 million from the Department of Trade and Industry as regional selective assistance

2. £15 million from English Partnerships in Property to ease further expansion of the Castle Bromwich plant

3. £17 million from local publically funded agencies to provide training and environmental improvement

Global inequalities

GROSS NATIONAL PRODUCT (GNP) PER CAPITA, 1993

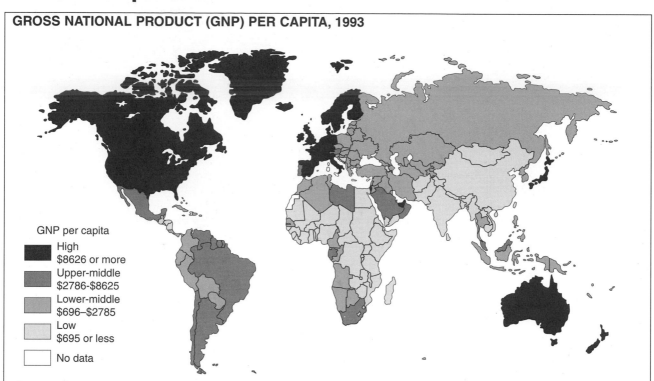

GNP per capita

- High $8626 or more
- Upper-middle $2786–$8625
- Lower-middle $696–$2785
- Low $695 or less
- No data

The map of GNP per capita shows a clear bias towards the **Economically More Developed Countries** (EMDCs). Western Europe, North America, Japan, and Australia come out on top according to GNP per capita: the highest values are Switzerland ($36,410), Luxembourg ($35,850), and Japan ($31,450). Approximately 15% of the world's population live in areas with a **high GNP per capita**. By contrast, 56% of the world's population live in areas classified as having a **low GNP per capita**. A number of countries have a GNP per capita of less than $200 per year: Rwanda, Burundi, Ethiopia, Tanzania, Uganda, Mozambique, Sierra Leone, and Vietnam.

PURCHASING POWER PARITY PER CAPITA, 1993

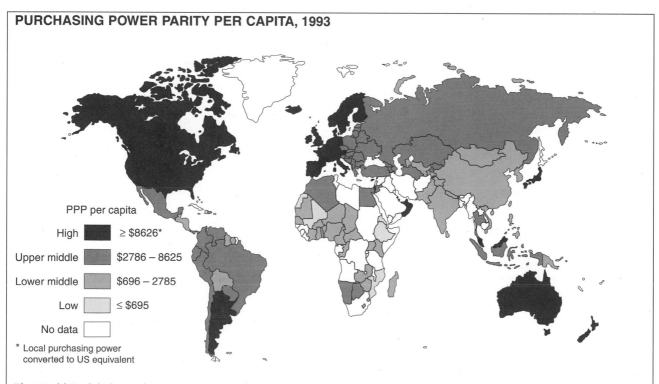

PPP per capita

- High ≥ $8626*
- Upper middle $2786 – 8625
- Lower middle $696 – 2785
- Low ≤ $695
- No data

* Local purchasing power converted to US equivalent

The World Bank believes that GNP per capita figures give a better indication of relative standards of living when converted into **purchasing power parity** (PPP). PPP relates average earnings to the ability to buy goods, i.e. how much you can buy for your money. For example, although wages in India are low compared to British wages, they can buy similar amounts of goods and services because local prices are also lower. In Switzerland, although GNP per head is very high, the high cost of goods and services lowers PPP. PPP lifts GNP per capita for most developing countries and former communist states but lowers it for developed countries.

HUMAN DEVELOPMENT INDEX (HDI)

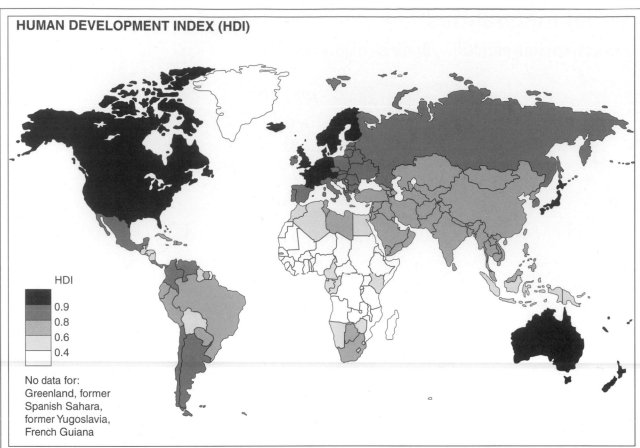

HDI

0.9
0.8
0.6
0.4

No data for:
Greenland, former
Spanish Sahara,
former Yugoslavia,
French Guiana

Since 1980 the United Nations (UN) has urged the use of the HDI as a measure of development. It is a more reliable and comprehensive measure of human development and well-being than GNP/head. It includes three basic components of human development:

- longevity (life expectancy)
- knowledge (adult literacy and average number of years schooling)
- standard of living (purchasing power adjusted to local cost of living)

The 1994 HDIs show Canada as the top HDI country, closely followed by Switzerland, Japan, and Sweden. The UK was placed 10th, ahead of Germany (11th), Ireland (21st), and Italy (22nd), but behind France (6th). At the other end, Guinea, Burkina Faso, and Afghanistan had the lowest HDI scores. Some countries, notably Saudi Arabia, Namibia, and the United Arab Emirates, had a higher GNP rank than HDI rank, suggesting scope to transfer oil or mineral revenue into human welfare projects.

National averages can conceal a great deal of information. HDIs can be created to show regional and racial variations, as shown in these diagrams.

REGIONAL AND RACIAL VARIATIONS IN THE HDI

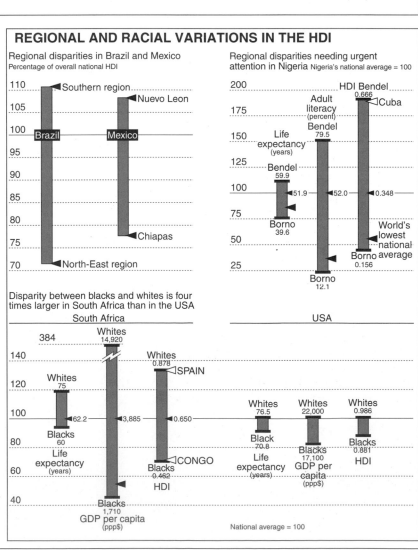

Regional disparities in Brazil and Mexico
Percentage of overall national HDI

Regional disparities needing urgent attention in Nigeria Nigeria's national average = 100

Disparity between blacks and whites is four times larger in South Africa than in the USA

National average = 100

Explaining inequalities in development

CLARK'S SECTOR MODEL

All developed countries have progressed from agricultural societies to industrial and service economies. For some, such as the UK, the transition was early, mostly in the nineteenth century, whereas for others, such as Ireland, it occurred during the twentieth century. Clark's model clearly shows the **transition** from an economy dominated by the primary sector to one dominated in turn by the secondary and tertiary sector. Change occurs because success in one sector produces a surplus revenue which is invested into new industries and technologies, thereby increasing the range of industries in an area. For example, in the UK, the cotton industry encouraged textile machinery, other metallurgical industries, and service industries. The sector model is descriptive and offers only a crude level of analysis. It does not say how or why a country developed, nor does it show internal variations within a country.

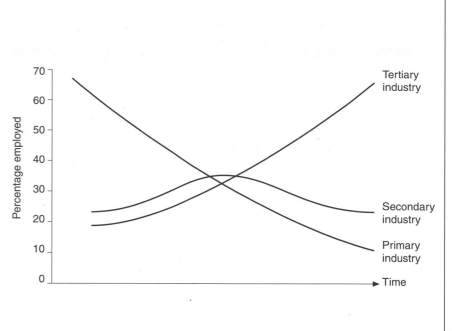

ROSTOW'S MODEL OF DEVELOPMENT

W. W. Rostow, a US economist, envisaged five stages in the development of an economy. His model is a useful starting point in describing and understanding levels of development. These levels can be described as:

1 **Traditional subsistence economy:** agricultural basis, little manufacturing, few external links, and low levels of population growth (stage 1 of the demographic transition model (DTM)). This stage is no longer present in the developed world.

2 **Preconditions for take-off:** external links are developed; resources are increasingly exploited, often by colonial countries or by multinational companies (MNCs); the country begins to develop an urban system (often with primate cities), a transport infrastructure, and inequality between the growing core and the underdeveloped periphery. The population continues to increase (stage 2 of the DTM). Again, this level has disappeared from developed countries.

3 **Take-off to maturity (sustained growth):** the economy expands rapidly, especially manufacturing exports. Regional inequalities intensify because of multiplier effects. This growth can be 'natural' (as in the case of most countries of the developed world), 'forced' (as in the former socialist countries of Eastern Europe), or planned (as in the Newly Industrialising Countries (NICs)).

4 **The drive to maturity:** diversification of the economy, and the development of the service industry (health, education, welfare, and so on). Growth spreads to other sectors and to other regions in the country. Population growth begins to slow down and stabilise (late stage 3 or early stage 4 of the DTM). Ireland, Greece, Spain, and Portugal are at this level.

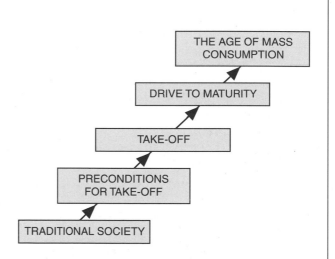

5 **The age of high-mass consumption:** advanced urban-industrial systems, with high production and consumption of consumer goods, such as televisions, compact disc players, dishwashers, and so on. Population growth slows considerably (stage 4 of the DTM). The UK and Germany characterise this level.

The main weaknesses of Rostow's model are:
- it is **anglo-centric**, based on the experience of North America and Western Europe
- it is **aspatial** and does not look at variations within a country. For example, within the UK, there are great disparities in the levels of development between the North and South - Rostow's model fails to pick this out.

MYRDAL'S MODEL OF CUMULATIVE CAUSATION

The core-periphery model, based on the work of **Gunnar Myrdal**, adopts a **spatial** outcome. It is seen as more useful than Rostow's model. In *Rich Lands and Poor Lands* (1957), Myrdal argued that, over time, economic forces increase regional inequalities rather than reduce them. He believed that development was caused by:

- Initial **comparative advantages**, e.g. resources such as location, minerals, or labour. These create the initial stimulus for an industry to develop in a particular location. In turn, a process of **cumulative causation** (multiplier effect) occurs as **acquired advantages**, such as improvements in infrastructure, skilled workforce, and increased tax revenues, are developed and reinforce the area's reputation, thereby attracting further investment, ensuring that it grows and stays ahead of other regions.

- Increased **spatial interaction**, i.e. skilled workers, investment, new technologies, and new developments gravitate to the growing area, the **core**, while the peripheral areas are inundated by manufactured goods from the core (the **backwash effect**), preventing the development of a local manufacturing base. As the core expands it may stimulate surrounding areas to develop due to increased consumer demand (the **spread effect**).

Three main stages can be identified in Myrdal's model:
- a traditional, pre-industrial stage, with few regional disparities (Rostow's stage 1)
- a stage of increased disparities caused by multiplier and backwash effects as the country industrialises (Rostow's stages 2 and 3)
- a stage of reduced regional inequalities as spread effects occur (Rostow's stages 4 and 5)

Myrdal's ideas have been used extensively in regional planning. In particular, they have been used in growth pole policies: places or districts favoured by their location, resources, labour, or market access are economically more attractive and are therefore developed by planners to form natural **growth poles**, expanding faster than other districts. Generally these are urban-industrial complexes which have good transport and accessibility, e.g. Dunkirk and Marseilles-Fos in France and Taranto in the Mezzogiorno, southern Italy.

Labour migration

Investment

GROWTH POLE THEORY

A growth pole is a dynamic and highly integrated set of industries organised around a leading industry or industrial sector. It is capable of rapid growth and generating multiplier effects or spillover effects into the local economy. The idea was originated by **Perroux** (1955) and developed by **Boudeville** (1966). It has been widely used in regional and national planning as a means of regenerating an area. Growth poles can, however, increase regional inequalities by concentrating resources in favoured locations.

FRIEDMANN'S STAGES OF GROWTH

1 **Pre-industrial economy:** independent local centres, no hierarchy. Similar to Rostow's stage 1.

2 **Transitional economy:** a single strong centre emerges. This dominates the colonial society as the stage of pre-conditions begins. A growing manufacturing sector encourages concentration of investment in only a few centres - hence a core emerges with a primate city.

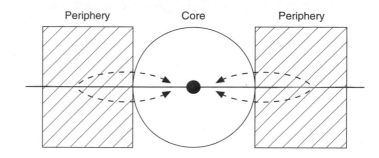

3 **Industrial economy:** a single national centre, strong peripheral sub-centres, increased regional inequalities between core and periphery; upward spiral in the core, downward spiral in the periphery (Myrdal's cumulative causation). In time, as the economy expands, more balanced national development occurs - sub-centres develop, forming a more integrated national urban hierarchy.

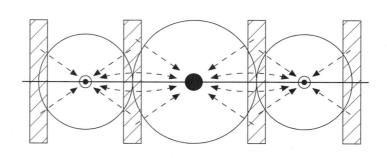

4 **Post-industrial economy:** a functionally interdependent urban system; the periphery is eliminated.

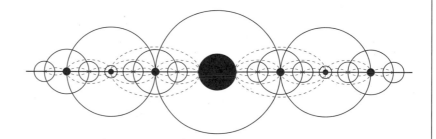

Friedmann believed that stage 4 has been reached in the USA, although there are still peripheral areas such as the Ozarks, Appalachians, and Alaska.

DEPENDENCY THEORY

According to **dependency theory**, countries become more dependent upon more powerful countries, frequently colonial powers, as a result of interaction and 'development'. As the more powerful country exploits the resources of its weaker colony, the colony becomes dependent upon the stronger power. Goods flow from the colony to support consumers in the overseas country.

Andre Frank (1971) described the effect of capitalist development on many countries as **'the development of underdevelopment'**. The problem of poor countries is not that they lack the resources, technical know-how, modern institutions, or cultural developments that lead to development, but that they are being exploited by capitalist countries.

Dependency theory is a very different approach to most models of development:

- it incorporates politics and economics in its explanation
- it takes into account the historical processes of how underdevelopment came about, i.e. how capitalist development began in one part of the world and then expanded into other areas (imperial expansion)
- it sees development as a revolutionary break, a clash of interests between ruling classes (the bourgeoisie) and the working classes (the proletariat)
- it stresses that to be developed is to be self-reliant and in control of national resources
- it believes that modernisation does not necessarily mean westernisation, and that underdeveloped countries must set goals of their own, appropriate to their own resources, needs, and values

Newly industrialising countries (NICs)

CHARACTERISTICS

An NIC is characterised by:

- significant average annual growth in manufacturing production (4.5% in Portugal, over 15% in South Korea)
- an increasing share of world manufacturing output
- a significant growth in (manufactured) export production
- an increasing proportion of the workforce in manufacturing industries
- a significant increase in GDP provided by manufacturing

Multinational companies (MNCs) play a part in NIC development. They facilitate economic development through their economic and political power and through their access to capital, skills, and knowledge. They also influence the type, scale, and location of manufacturing.

Three main groups of NICs have been identified:

- 'Asian tigers' such as Hong Kong, Singapore, South Korea, and Taiwan
- Latin American NICs such as Brazil and Mexico
- European NICs - Spain, Portugal, Greece, and the former Yugoslavia

Some of these NICs have been redefined, e.g. Spain was classified as an EMDC in 1983.

STAGES IN THE EMERGENCE OF AN NIC

1 **Traditional society**
   Labour intensive industries, low levels of technology. Local raw materials - food processing and textiles common.

2 **Import substitution industries (ISIs)**
   Reduction of expensive imports by development of home industries. Protectionist policies, e.g. high trade tariffs on manufactured goods and in the car industry.

3 **Export-orientated industries (EOIs)**
   High-technology, capital intensive industries. R&D functions. Rapid growth and development.

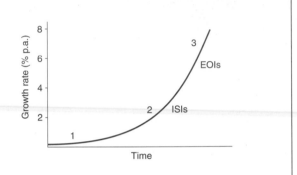

THE EFFECTS OF RAPID INDUSTRIALISATION

Economic effects

- Raises standards of living.
- Benefits the government rather than the people.
- Change in labour force:
  - (i) shift to manufacturing (secondary) industries, especially of young working population
  - (ii) decline in agricultural employment and productivity; growth of rural unemployment
  - (iii) increase in service industries, e.g. transport, retailing, and administration
- Growth of urban infrastructure.
- Increased competition for land, raising land prices.

Social effects

Rural areas

- Social imbalance between affluent industrial minority/commercial workers and the rest of the population.
- Population change, e.g. age-selective migration to urban areas.
- Decline of traditional values and lifestyles.
- Increased dependence on remittances from urban workers.
- Poor welfare systems.

Urban areas

- In-migration leads to rapid development of shanty towns, increased birth rates, and population growth.
- Concentration of unemployed and poor in shanty towns.
- High rates of crime, illiteracy, disease, and so on in areas of impoverished housing.
- Low levels of service provision.
- Unsatisfactory working conditions (sweatshops).
- Political and social unrest.

Environmental effects

- Resource exploitation can damage the natural environment and can lead to the destruction of whole habitats.
- Rivers polluted by industrial waste.
- Air pollution, e.g. in Taipei, Taiwan.
- Unsafe working practises may lead to environmental disasters, e.g. at Bhopal, India, in 1985, toxic gas from the Union Carbide factory lead to widespread blindness in the area.
- Urban blight, e.g. derelict buildings, contaminated land.
- Limited environmental legislation.

Tourism

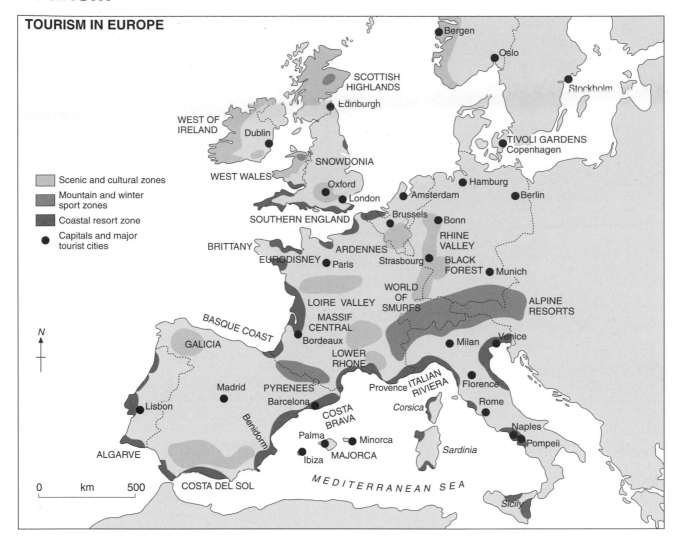

TOURISM IN EUROPE

Legend:
- Scenic and cultural zones
- Mountain and winter sport zones
- Coastal resort zone
- Capitals and major tourist cities

Map labels: Bergen, Oslo, Stockholm, SCOTTISH HIGHLANDS, Edinburgh, WEST OF IRELAND, Dublin, SNOWDONIA, WEST WALES, TIVOLI GARDENS Copenhagen, Hamburg, Oxford, London, Amsterdam, Berlin, SOUTHERN ENGLAND, Brussels, Bonn, RHINE VALLEY, BRITTANY, ARDENNES, Strasbourg, BLACK FOREST, Munich, EURODISNEY, Paris, WORLD OF SMURFS, ALPINE RESORTS, LOIRE VALLEY, MASSIF CENTRAL, BASQUE COAST, GALICIA, Bordeaux, LOWER RHONE, Milan, Venice, Madrid, PYRENEES, Provence, ITALIAN RIVIERA, Florence, Lisbon, Barcelona, COSTA BRAVA, Corsica, Rome, Benidorm, Palma, Minorca, Naples, Pompeii, Ibiza, MAJORCA, Sardinia, ALGARVE, COSTA DEL SOL, Sicily, MEDITERRANEAN SEA

Scale: 0 km 500

N

The tourist industry is now one of the world's largest industries. The number of foreign holidays rose from approximately 25 million in 1950 to over 350 million in the mid-1990s. The number of holidays at home is even greater. The rise in tourism is related to a number of economic and social trends:

- increased leisure time
- cheaper, faster forms of transport, especially air travel
- an increase in disposable income and a broadening of lifestyle expectations
- the growth of the package holiday
- greater media exposure, travel programmes on television, and so on
- a rise in the number of second holidays, short breaks, weekend trips, and so on

The main destinations in Europe can be divided into six broad, overlapping types of destination:
1 **Coastal**, e.g. the Costa del Sol, Blackpool, the Algarve
2 **Scenic** landscapes, e.g. South-West Ireland, Scotland
3 **Mountain and ski** centres, e.g. the Alpine resorts
4 **Capital cities and heritage** centres, e.g. London, Paris, Bruges, Oxford
5 **Leisure** centres, e.g. EuroDisney, Tivoli Gardens, Center Parcs
6 **Business-conference** centres, e.g. Korpilampi near Helsinki

The rise in tourism has had a serious effect on many countries - not just economically but in terms of social impact and environmental damage. Although most of the global pattern of tourism is between developed countries, many developing countries are turning to tourism as a way to develop and as a means of obtaining foreign income.

TOURISM IN SOUTH AFRICA

Tourists are attracted to developing countries, such as South Africa, Kenya, and Sierra Leone, for a number of reasons. South Africa, in particular, for example, is rapidly becoming a popular destination for foreign visitors. The reasons for this include:

- its rich and varied **wildlife** and world-famous **game reserves**, e.g. the Kruger National Park
- a **warm and sunny** climate, especially in December and January
- glorious **beaches** in Natal
- the **cultural heritage** and tradition of the Zulu, Xhosa, and Sotho peoples
- it is relatively **cheap** compared to developed countries
- **English** is widely spoken and there are many links with the UK
- it is perceived as a **safe destination** since the collapse of apartheid and the election of the new ANC government

There are a number of **benefits** that tourism can bring to a developing economy such as that of South Africa:

1. **Foreign currency**: the number of foreign tourists has increased by 15% per annum during the 1990s and contributes 3.2% of South Africa's GDP.
2. **Employment**: thousands are employed in formal (registered) and informal (unregistered) occupations ranging from hotels and tour operators to cleaners, gardeners, and souvenir hawkers.
3. It is a more **profitable** way to use semi-arid grassland: estimates of the annual returns per hectare range from R60-80 for pastoralism to R250 for dry-land farming and R1000 for game parks and tourism.
4. **Investment**: over R5 billion was invested in South Africa's tourist infrastructure in 1995, upgrading hotels, airlines, car rental fleets, roads, and so on.

However, there are a number of **problems** that have arisen as a result of the tourist industry:

1. There is undue **pressure** on natural ecosystems, leading to soil erosion, litter pollution, and declining animal numbers.
2. Much tourist-related employment is **unskilled**, **seasonal**, **part-time**, **poorly paid** and lacking any rights for the workers.
3. **Resources** are spent on providing for tourists while local people may have to go without.
4. A large proportion of **profits** goes to overseas companies, tour operators, and hotel chains.
5. **Crime** is increasingly directed at tourists; much is petty crime but there have been some very serious incidents, such as rape.
6. Tourism is very **unpredictable**, varying with the strength of the economy, cost, safety, alternative opportunities, stage in the family life-cycle, and so on.

ECOTOURISM

Ecotourism is a specialised form of tourism where people want to see and experience relatively untouched natural environments. Typical destinations include game reserves, coral reefs, mountains, and forests. A classic example would be gorilla-watching in Central Africa. It has recently been widened to include 'primitive' indigenous people. It is widely perceived as an 'acceptable' type of tourism and a form of sustainable development, i.e. a type of tourism that can be developed without any ill-effects on the natural environment. However, much that passes for ecotourism is merely an expensive package holiday cleverly marketed with the 'eco' label.

Ecotourism developed as a form of specialised, flexible, low-density tourism. It emerged because mass tourism was seen as having a negative impact upon natural and social environments, and because there was a growing number of wealthy tourists dissatisfied with package holidays. In the original sense, people who took an 'ecotourism' holiday were prepared to accept quite simple accommodation and facilities. This is sustainable and has little effect upon the environment. However, as a location becomes more popular and is marketed more, the number of tourists increases resulting in more accommodation and improved facilities, e.g. more hotels with showers and baths, air conditioning, bars, sewage facilities, and so on. This leads to an unsustainable form of ecotourism as it destroys part of the environment and/or culture that the visitors want to experience.

Environmental issues

A number of issues were discussed at the Earth Summit in Rio de Janeiro in 1992. The debate rests not just on the issues themselves but on the relationship between the developed and the developing world and the idea of sustainable development in both the North and the South.

OZONE DEPLETION

- international response to the ozone threat is held up as an example of how effective global agreements can be
- in 1987, 27 countries signed the Montreal Protocol – an accord to curb the use of chlorofluorocarbons

EXTINCTIONS

- humanity is driving species extinction at 25,000 times the natural rate
- species in the oceans and tropical rain-forests may yield medicines and other potentially valuable commodities
- species decline also has a moral element – the argument that species should be preserved for future generations

EROSION

- each year the world loses 24 billion tons of topsoil
- processes of soil degradation and desertification affect both the developed and developing world
- not only affects fertility but can also silt up rivers and dams

THE STATE OF THE PLANET

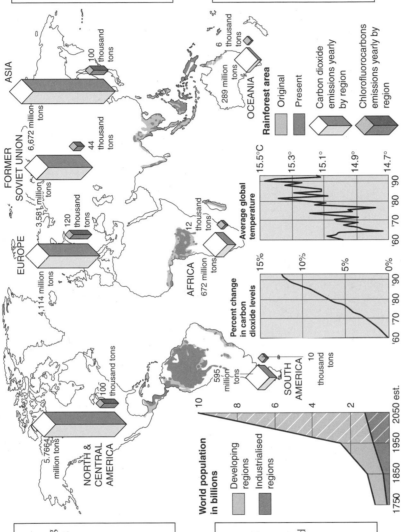

ASIA 6,672 million tons / 100 thousand tons

FORMER SOVIET UNION 3,581 million tons / 44 thousand tons

EUROPE 4,114 million tons / 120 thousand tons

AFRICA 672 million tons / 12 thousand tons

OCEANIA 289 million tons / 6 thousand tons

NORTH & CENTRAL AMERICA 5,7664 million tons / 100 thousand tons

SOUTH AMERICA 595 million tons / 10 thousand tons

Rainforest area
Original
Present

Carbon dioxide emissions yearly by region

Chlorofluorocarbons emissions yearly by region

Average global temperature
15.5°C
15.3°
15.1°
14.9°
14.7°
'60 '70 '80 '90

Percent change in carbon dioxide levels
15%
10%
5%
0%
'60 '70 '80 '90

World population in billions
10
8
6
4
2
1750 1850 1950 2050 est.
Developing regions
Industrialised regions

HAZARDOUS WASTE

- most chemical wastes are produced by the industrialised nations
- many of the chemicals are shipped to the developing countries for cheap disposal
- many of these chemicals have not been tested for potential effects on human health and the environment

DIRTY DRINKING WATER

- more than 1.2 billion people lack safe water to drink and over 1.8 billion do not have adequate sanitation
- clean water would save the lives of two million children under five each year
- every year waterborne diseases cost India 73 million working days

OVERPOPULATION

- each year nearly 100 million people are added to the earth's population, meaning that it will double in just over forty years
- increasing population puts pressure on resources and increases environmental stress
- consensus that women should be given greater control over reproduction and better access to family planning

CLIMATE CHANGE

- carbon dioxide (CO_2) released by burning oil, gas, and coal and by forest burning represents half the greenhouse gases
- chlorofluorocarbons from aerosols, food packaging, and fridges are responsible for another quarter
- most of the rest comes from methane from cattle

DEFORESTATION

- every week at least one million acres of forest are cleared or degraded worldwide
- problems include soil degradation, species destruction, and increased flooding
- suggestions that the huge debts of countries like Brazil should be 'traded' for reduced destruction of the rainforest

Resource exploitation

All environmental issues are essentially about the use and distribution of resources.

A CLASSIFICATION OF NATURAL RESOURCES

Stock resources can be divided into:

- those 'consumed by use', e.g. fossil fuels or phosphate fertilisers

- those that are 'recyclable', e.g. metals which can be recovered

The **issue** is the prospect of 'running out' of non-renewable resources. This has been a significant concern since the 1960s and 1970s.

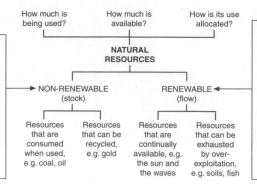

How much is being used? How much is available? How is its use allocated?

NATURAL RESOURCES

NON-RENEWABLE (stock)   RENEWABLE (flow)

Resources that are consumed when used, e.g. coal, oil

Resources that can be recycled, e.g. gold

Resources that are continually available, e.g. the sun and the waves

Resources that can be exhausted by over-exploitation, e.g. soils, fish

Flow resources can be divided into:

- those 'not affected by human action', e.g. solar radiation, tidal energy

- those 'affected by human use', e.g. soils, forest ecosystems

The **issue** is the danger of irreversible damage to the biosphere. This became a significant concern in the 1980s and 1990s.

THE RESOURCE BASE

Resource scarcity would lead to higher prices, conservation, recycling, or substitution.

Proven reserves
Can be extracted at current prices with current technology.

Conditional reserves
Not economic under prevailing conditions of technology and economics.

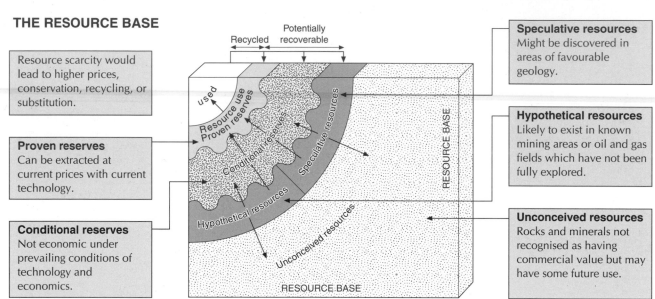

Recycled   Potentially recoverable

used

Resource use
Proven reserves

Conditional reserves

Speculative resources

Hypothetical resources

Unconceived resources

RESOURCE BASE

RESOURCE BASE

Speculative resources
Might be discovered in areas of favourable geology.

Hypothetical resources
Likely to exist in known mining areas or oil and gas fields which have not been fully explored.

Unconceived resources
Rocks and minerals not recognised as having commercial value but may have some future use.

ENVIRONMENTAL THEMES

Uneven distribution of costs and benefits over time

- issues of 'intergenerational equity'

- are we using up resources which should be saved for future generations?

- are we storing up problems that must be faced in the future?

Quantifying the unquantifiable

- intangible nature of environmental costs and benefits

- 'environmental value' cannot readily be costed in monetary terms

- activities which affect the environment may bring jobs, profit, cheaper power, faster journeys, and so on

THEMES

Externalities

- costs and benefits which flow from environmental actions are distributed unevenly in space and between different social groups

- issues include the relocation of 'dirty' industries to the developing world - why is this happening, and should it happen?

Uncertainty

- all environmental issues have a degree of uncertainty

- this rests not only on what should be done (solutions) but also on causes

- it is often difficult to be sure about environmental damage

Deforestation of the tropical rainforest

Deforestation is the deliberate removal of forest for commercial reasons such as agriculture, fuelwood and charcoal, construction, logging, mining, and road building.

Loss of biodiversity
- forests are a source of potential foods, drinks, medicines, contraceptives, resins, scents, and pesticides

Vegetation change
- in some regions there is evidence that cleared areas are colonised by grasses rather than trees, resulting in the loss of forest

Soil erosion
- deforested slopes encourage surface runoff (the loss of the canopy reduces interception)
- the runoff erodes and removes soil via rills and gullies

ENVIRONMENTAL CONSEQUENCES

Laterisation of soils
- the process whereby the oxides of iron are leached and deposited
- the so-called 'hard pan' is then revealed by soil erosion
- its concrete-like hardness makes it impossible to farm

Floods
- increased runoff results in shorter lag times
- rapid movements of water over short time periods results in flooding

Climate change
- burning of forests for clearance increases the CO_2 in the atmosphere - this could be leading to global warming
- some argue that rainforests are the lungs of the earth, changing CO_2 into oxygen

These traditional views of environmental degradation are being increasingly questioned by scientists.

CASE STUDY: THE COLONISATION OF AMAZONIA

Modern developments have brought millions of new migrants into the Brazilian rainforest. Reasons include rubber tapping, logging, ranching, mining, and especially small-holding farming. Peasants have migrated from the crowded North-East for five reasons:

(i) increased mechanisation, (ii) the attraction of acquiring new land, (iii) the perception that Amazonia is fertile, (iv) the building of the Trans-Amazonia Highway, (v) government policies of colonisation.

Ranching
- cattle ranching by large commercial companies
- forest removal by bulldozing and firebombing
- ranching accounts for 40% of cleared land in Eastern Amazonia
- after four years land is often exhausted and abandoned

Smallholders
- peasant farmers survive by growing tree crops of nuts and fruit
- clearance by forest fires
- in 1987, 350,000 separate forest fires
- few farmers can develop beyond subsistence level

Logging
- 3.5 million cubic metres of mahogany was exported between 1971 and 1991
- rainforest will not regenerate if cleared in broad open swathes
- conservation is possible: in Puerto Rico, loggers take less than 1% of saleable stock and always leave divided trunks which do regenerate

Road building
- Manaus-Boa Vista Highway (BR-174) took land from Waimiri-Attroari tribe
- this affected the traditional sustainable use of the forest by the tribe:
  (i) they lost much land to actual road building
  (ii) the highway allowed settlers to move in and take more land

Percentage of state deforested
13.8 1978
1.8 1990
— Major road

Roraima *1.7* **0.1**

Amapa *1.3* **0.1**

Para *19.3* **1.3**

Amazonas *1.3* **0.1**

B R A Z I L

Mato Grosso *10.4* **2.5**

Acre *6.7* **1.6**

Rondônia *13.9* **1.8**

0 km 500

Government land grants and access roads completed during the period have accelerated deforestation in Rondônia and Acre. Note, however, the land as yet unaffected – 98.7 per cent in closely forested Amazonas.

Deforestation in Amazonia

Land degradation and desertification

THE 'FIVE Ds'

Desertification is land degradation in arid, semi-arid, and dry sub-humid areas resulting in the spread of desert-type conditions.

There are five elements to this process - the Five Ds:

(i) **Drylands** - 'susceptible to experiencing full desert conditions if mismanaged'; this is a *climatic* definition which implies fragility.

(ii) **Degradation** - 'a reduction or destruction of the biological potential'; this is usually associated with unwise *human* practices, e.g. overgrazing, deforestation, trampling, overproduction.

(iii) **Drought** - 'two or more years with rainfall substantially below the mean'; a *climatic* variable.

(iv) **Dessication** - 'the process of longer-term reductions in moisture availability resulting from a dry period at the scale of decades'; a *climatic* variable.

(v) **Desertification** - the combination of human and climatic variables which leads to the irreversible decline of the land.

IS DESERTIFICATION A MYTH?

There are elements of *uncertainty* to the debate about desertification.

Myth		Reality
Desertification affects one-third of the world's land area.		Such data are inaccurate, with problems of separating natural variations from human actions.
Drylands are fragile ecosystems that are highly susceptible to degradation and desertification.		Drylands are more resilient than first supposed and are well-adapted to cope with and respond to disturbance.
Desertification is a primary cause of human suffering and misery in drylands.		Social problems may be due to political mismanagement rather than drought and desertification.
Western aid bodies such as the UN are needed to solve the problem of desertification.		Much evidence that traditional human systems have evolved to cope with environmental disturbance.

The causes of desertification

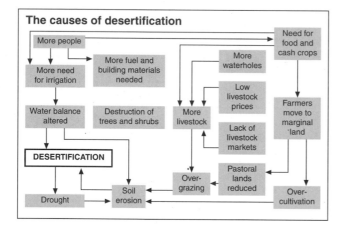

Desertification feedback cycles

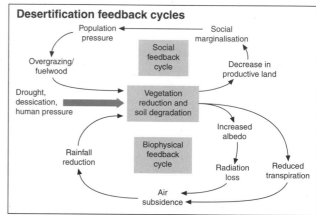

COMBATING DESERTIFICATION

Cause of desertifi-cation	Strategies for prevention	Problems and drawbacks
Over-cultivation	**Use of fertilisers**: these can double yields of grain crops, reducing the need to open up new land for farming. **New or improved crops**: many new crops or new varieties of traditional crops with high yielding and drought resistant qualities could be introduced. **Improved faring methods**: use of crop rotation, irrigation, and grain storage can all increase income and reduce pressure on land.	• Cost to farmers. • Artificial fertilisers may damage the soil. • Some crops need expensive fertilisers. Risk of crop failure. • Some methods require expensive technology and special skills.
Over-grazing	**Improved stock quality**: through vaccination programmes and the introduction of better breeds, yields of meat, wool, and milk can be increased without increasing the herd sizes. **Better management**: reducing herd sizes and grazing over wider areas would both reduce soil damage.	• Vaccination programmes improve survival rates, leading to bigger herds. • Population pressure often prevents these measures.
Deforesta-tion	**Agroforestry**: combines agriculture and forestry, allowing the farmer to continue cropping while using trees for fodder, fuel, and building timber. Trees protect, shade, and fertilise the soil. **Social forestry**: village-based tree-planting schemes involve all members of a community. **Alternative fuels**: oil, gas, and kerosene can be substituted for wood as sources of fuel.	• Long growth time before benefits of trees are realised. • Expensive irrigation and maintenance may be needed. • Expensive. Special equipment (e.g. stoves) required.

Acid rain

Acid rain or acid deposition is the people-made increase in acidity brought about by air pollution.

SOURCES
Fossil fuels (coal, oil, gas) are burned and release exhaust gases, SO_2 and NO_x, which react with sunlight and ozone and then dissolve to produce acid rain.

DISPERSION AND DEPOSITION

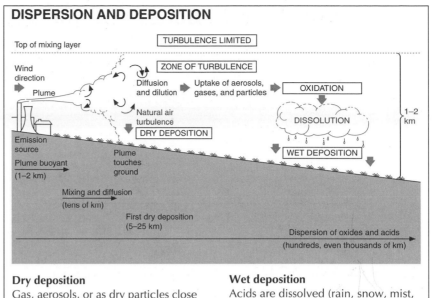

Dry deposition
Gas, aerosols, or as dry particles close to emission sources, resulting in urban and industrial damage to buildings and cultural treasures.

Wet deposition
Acids are dissolved (rain, snow, mist, hail) and fall back great distances from the sources, resulting in damage to lakes and forests.

EFFECTS OF ACID RAIN

Forests
Suggested process: release of toxic metals into the soil (especially aluminium), causing damage to tree roots and nutrient deficiency and disease; ability to cope with stress (frost, drought, pests) reduced.

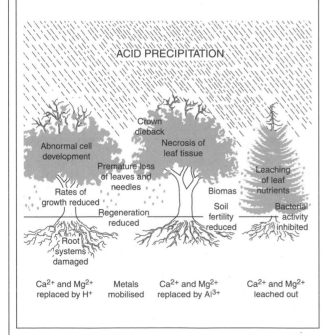

Lakes
Suggested process: disturbs the balance of salt intake for fish, clogging gills with sticky mucus; in south Norway all lakes in an area covering 1.3 million hectares are devoid of fish.

Solutions: *Prevention or cure*

Broadly speaking, environmentalists and 'victims' stress the ecological risks; 'polluters' emphasise the uncertainties.

- reduce acidity by adding lime to lakes, soils, and rivers, but this is costly and temporary; lime may also have problems associated with the sudden surge of alkalinity or a rise in water temperature due to the exothermic reaction

- prevention by reducing SO_2 and NO_x emissions, especially from power stations

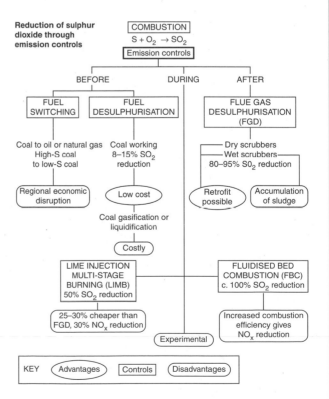

The greenhouse effect and global warming

THE GREENHOUSE EFFECT

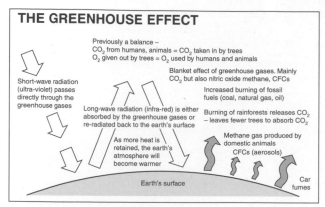

Previously a balance –
CO_2 from humans, animals = CO_2 taken in by trees
O_2 given out by trees = O_2 used by humans and animals

Short-wave radiation (ultra-violet) passes directly through the greenhouse gases

Blanket effect of greenhouse gases. Mainly CO_2 but also nitric oxide methane, CFCs

Increased burning of fossil fuels (coal, natural gas, oil)

Long-wave radiation (infra-red) is either absorbed by the greenhouse gases or re-radiated back to the earth's surface

Burning of rainforests releases CO_2 – leaves fewer trees to absorb CO_2

As more heat is retained, the earth's atmosphere will become warmer

Methane gas produced by domestic animals

CFCs (aerosols)

Earth's surface

Car fumes

GLOBAL WARMING AND POSSIBLE EFFECTS ON THE PHYSICAL ENVIRONMENT

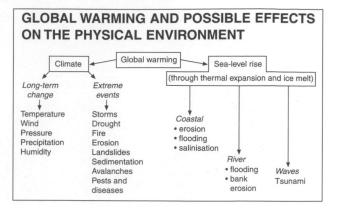

Climate ← Global warming → Sea-level rise

(through thermal expansion and ice melt)

Long-term change
Temperature
Wind
Pressure
Precipitation
Humidity

Extreme events
Storms
Drought
Fire
Erosion
Landslides
Sedimentation
Avalanches
Pests and diseases

Coastal
• erosion
• flooding
• salinisation

River
• flooding
• bank erosion

Waves
Tsunami

NEGATIVE EFFECTS: SEA-LEVEL RISE IN BANGLADESH

Probably the most important geomorphologic consequence of global warming would be a worldwide rise in sea-level due to the thermal expansion of the upper layers of the oceans and the melting of land ice.

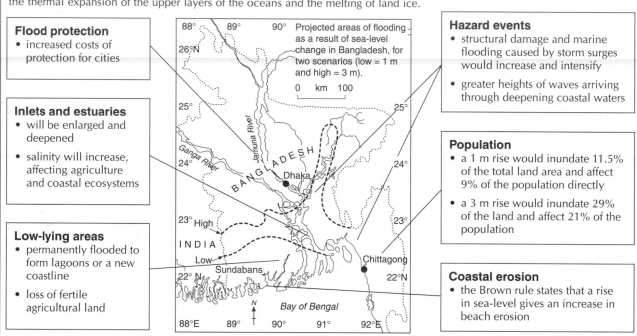

Flood protection
• increased costs of protection for cities

Inlets and estuaries
• will be enlarged and deepened
• salinity will increase, affecting agriculture and coastal ecosystems

Low-lying areas
• permanently flooded to form lagoons or a new coastline
• loss of fertile agricultural land

Projected areas of flooding as a result of sea-level change in Bangladesh, for two scenarios (low = 1 m and high = 3 m).

Hazard events
• structural damage and marine flooding caused by storm surges would increase and intensify
• greater heights of waves arriving through deepening coastal waters

Population
• a 1 m rise would inundate 11.5% of the total land area and affect 9% of the population directly
• a 3 m rise would inundate 29% of the land and affect 21% of the population

Coastal erosion
• the Brown rule states that a rise in sea-level gives an increase in beach erosion

IS GLOBAL WARMING ALWAYS BAD? EASTERN AND CENTRAL CANADA

There are potential benefits from global warming for some countries in high latitudes. There will be a movement rather than a reduction of agricultural belts and increasing CO_2 may stimulate plant growth. In Canada, global warming would have socio-economic effects on agriculture, recreation, forestry, and shipping.

Agriculture
• mostly positive effects
• crops would increase productivity with more CO_2
• arable areas would expand northwards
• growing season would increase
• chances of drought would increase

Forestry
• reduction of the boreal forest area by as much as 20% in Quebec
• however, this would be offset by increased intensity of growing season (biomass increase of as much as 50%)
• the shift would lead to the decline of the pulp and paper towns from Quebec to Alberta

Shipping
• year-round shipping on Great Lakes
• reduced ice cover on Hudson Bay

Recreation
• extended summer season would allow expansion of golf, hiking, camping, and watersports
• decrease in revenue from the skiing industry with a decline in the number of skiable days

INDEX